# Executive summary

It is widely agreed that the world's climate is changing and will continue to change for at least many decades into the future. This will affect climatic conditions in Britain and will require the UK construction industry to respond accordingly. Recent extreme weather conditions have caused significant damage to buildings, roads and railways. Most evident have been rainfall and flooding events. However the changing climate also poses other risks to construction project, including:

- delay to construction programmes
- poorer internal environment leading to mould growth in dwellings
- subsidence and heave
- slope instability
- damage to fabric of building, particular its effects on cladding
- structural damage from wind related events
- effect on roof drainage

At present there is little evidence that UK construction companies are implementing, or even considering, adaptation strategies or mitigation measures to reduce these impacts. Although climate change is beginning to be taken into account in the development of building standards and construction guidance, overt references to the phenomenon are still lacking from much of the current guidance, despite the likely effects falling within most project life-cycles. This makes it hard for construction professionals to know when additional considerations should be applied in decision-making.

The book describes a method for assessing the risk that should help designers and constructors to make rational decisions about whether to incorporate climate change consequences in their projects. The method suggested follows the generic approach for risk decision-making, so it can be adapted to complement existing internal decision-making processes.

It is important to recognise that knowledge of climate change will continue to increase as changes to the UK climate are seen and the models used by the scientists are refined. Construction professionals therefore need to be aware that the likelihood and magnitude of climate change events may change over the coming decades and that up-to-date figures should be used in any risk assessment processes.

CIRIA C638                                     **London, 2005**

# Climate change risks in building – an introduction

**S Vivian**           URS Corporation

**N Williams**        URS Corporation

**W Rogers**          URS Corporation

**CIRIA** *sharing knowledge* ■ *building best practice*

Classic House, 174–180 Old Street, London EC1V 9BP, UK
TEL  +44 (0)20 7549 3300       FAX  +44 (0)20 7253 0523
EMAIL  enquiries@ciria.org
WEBSITE  www.ciria.org

# Summary

Rainfall and flooding have caused significant damage to buildings and transport infrastructure in the UK over the past few years. Recent extreme weather conditions have shown that all parts of the world are vulnerable to the consequences of the global climatic change, which, most people now accept, is taking place.

The effects of climate change range from delay to construction programmes and mould growth in dwellings to subsidence and heave, slope instability and damage to the building fabric.

There is little evidence that UK construction companies are implementing or considering adaptation strategies or mitigation measures to reduce these impacts. Construction clients (both public- and private-sector), engineers, contractors and other interest groups will therefore benefit from this up-to-date review of the implications of climate change and the practical guidance it contains on assessing and managing the associated risks, such as ground movement, rain penetration and wind loading.

**Climate change risks in building – an introduction**

Vivian, S; Williams, N; Rogers, W

*CIRIA*

CIRIA C638          © CIRIA 2005          RP668          ISBN 0-86017-638-X

**British Library Cataloguing in Publication Data**

A catalogue record is available for this book from the British Library.

| Keywords | | |
| --- | --- | --- |
| Climate change, concrete and structures, construction management, design and buildability, environmental good practice, facilities management, flooding, housing, materials, project management, risk and value management, site management, sustainable construction. | | |
| **Reader interest** | **Classification** | |
| Engineers, consultants, clients, local authorities | AVAILABILITY | unrestricted |
| | CONTENT | guidance document |
| | STATUS | committee-guided |
| | USER | engineers, consultants, clients, local authorities |

Published by CIRIA, Classic House, 174–180 Old Street, London EC1V 9BP, UK.

# Acknowledgements

**Research contractor**

The research contractor for CIRIA Research Project 668, "Technical risk assessment of climate change for the construction industry" was URS Corporation. The principal contributors from URS Corporation were Sally Vivian, Nicole Williams and Will Rogers.

**Authors**

**Sally Vivian**

Sally Vivian is a principal environmental scientist with more than 14 years' experience performing a wide range of environmental and broader sustainability services to industry and public bodies. She has particular expertise in identifying and managing environmental business risks and has spoken at conferences and events on the application of environmental and broader sustainability management systems within the construction sector.

**Will Rogers**  MA CEng MICE

Mr Rogers is a technical director at URS with a background in highway engineering and project management. He is involved in the environmental impact assessment of highway schemes and private developments, and has particular expertise in the impacts of such proposals on the water environment.

**Nicole Williams**

Ms Williams has over eight years' experience in environmental management particularly in relation to research, policy development and application and management systems. As a consultant she has had significant experience in identifying and understanding environmental risks affecting industry and reviewing the effectiveness of control systems developed to manage and mitigate the impacts.

**Project steering group**

CIRIA wishes to express its thanks to the members of the project steering group for their contributions to the project:

| | |
|---|---|
| Maurice Brown | Entec |
| Tara Butler | RICS |
| Richenda Connell | UKCIP |
| Gavin Davies | Arup and Partners |
| George Fordyce | NHBC |
| Sian Lewis | House Builders Federation |
| Mark Goldthorpe | South East Climate Change Partnerships |
| George Henderson | WS Atkins (for DTI) |
| Geoff Levermore | UMIST |
| Sam Evan | The Housing Corporation |
| Jane Milne | Association of British Insurers |
| Brian Neale | HSE |
| Steve Robert | Marsh |
| Janet Young | The Peabody Trust. |

**Corresponding member**  Alan Yates  BRE

**CIRIA manager**

CIRIA's research manager for this project was Joanne Kwan, who also chaired the steering group.

**Funders**

The project was funded by the Department of Trade and Industry (through the Partners In Innovation programme), the National House Building Council, the Housing Corporation and the Royal Institution of Chartered Surveyors.

CIRIA and the authors gratefully acknowledge the support of these funding organisations and the technical help and advice provided by the members of the steering group. Contributions do not imply that individual funders necessarily endorse all views expressed in published outputs.

CIRIA C638

# Contents

## Figures

## Tables

# 1 Introduction

This book aims to raise the awareness of professionals in the construction industry of the possible impact of climate change on the industry and to provide assistance in the identification and management of risks associated with climate change.

## 1.1 SCENE SETTING

### 1.1.1 Background

It is widely agreed that everyday human activities, in particular the reliance on fossil fuels for the generation of energy and for transport, have contributed to the increase in the amount of greenhouse gases released to the atmosphere. This increase has caused, or at least contributed significantly to, the observed changes in global weather conditions known as *global warming* or *climate change*.

---

**POINT TO NOTE**

**Greenhouse gases**

Carbon dioxide (from energy production, transport etc), methane (eg from agriculture, landfill) and hydrofluorocarbons (eg from refrigeration) are all examples of greenhouse gases. These gases trap energy in the lower atmosphere and result in a warming of the global climate, with consequent changes in weather patterns such as increased rainfall intensity and storm frequency (Houghton *et al*, 2001).

---

The world's climate is likely to continue changing despite efforts to reduce the level of greenhouse gases released to the atmosphere in the future. It is generally accepted that climate change predictions for the next 30–40 years are relatively certain as they are based on current greenhouse gas accumulations. Beyond this timescale the predictions are likely to vary depending on the success in reducing or limiting emissions. In terms of the effect on the built environment, many observers consider that some aspects of climate change are already being seen, such as the increased numbers of storms and flooding in recent years.

Built environment stock in the UK is generally expected to last between 50 and 100 years. Roads, railways and other infrastructure are designed to last even longer. It is important for the construction industry to recognise that there will be an expectation among users and the wider community that developments designed now will be proof against climate changes within the lifetime of the development.

The industry needs to understand how to design projects to be proof against future climate change while facing uncertainty over the possible climate conditions 50 years or more from now. If the risks associated with climate change are not factored into designs there are likely to be future implications, such as:

- higher maintenance costs
- costs relating to contractual arrangements with buildings and infrastructure not meeting the design life
- higher insurance premiums and/or difficulties in gaining insurance
- indirect costs associated with reputation and public perception.

Recent extreme weather conditions have caused significant damage to buildings, roads and railways in the UK. Most evident have been rainfall and flooding events. In April 1998, severe flooding occurred in the Nene and Great Ouse valleys, directly affecting more than 10 000 people in Northampton. The winter of 2000/2001 was one of the wettest on record. In 2003 a range of extreme events was recorded in the UK (see also Box 2.1).

These events are indicative of the sorts of factors that the industry will have to take into account in future design and construction. This book provides UK-focused guidance that will allow the industry to assess and manage the risks involved in this challenge.

> **POINT TO NOTE**
>
> For the purposes of this study and publication, the "construction industry" includes all the following activities:
>
> • planning, design, construction, maintenance and refurbishment of commercial and domestic buildings and infrastructure.
>
> The book is relevant to all involved in the above activities such as clients, contractors, designers, house-builders, and facilities managers.

### 1.1.2 Climate change activity at International level

The Intergovernmental Panel on Climate Change (IPCC) is the body responsible for assessing and disseminating climate change information at the global level. Reports produced by the IPCC are largely scientific and technical, although they do encompass discussions on sensitivities and adaptation techniques across a large variety of sectors including ecosystems, water management, agriculture and construction.

While the IPCC does not appear to have identified specific adaptation techniques for the construction industry, the cost to the global economy from flooding and other extreme events that damage buildings and infrastructure has been discussed and is estimated by IPCC to be in the order of billions of dollars (McCarthy *et al*, 2001).

### 1.1.3 Climate change activity at UK and regional level

There is much activity within the UK aiming to identify likely climate change impacts and consider potential adaptation strategies required at a national level. Table 1.1 identifies some of the organisations involved in these activities, their roles and sources of further information, such as key specific research projects. These include work being done by several organisations including the Energy Saving Trust and the Action Energy programme of the Carbon Trust to reduce the amount of energy used by increasing energy efficiency in buildings and thereby reducing the amount of greenhouse gases released to the atmosphere, in turn helping to reduce the extent of climate change in the future.

**Table 1.1** *Overview of organisations involved in climate change research*

| Organisation | Relevant role | Website |
|---|---|---|
| UK Climate Impacts Programme (UKCIP) | UK Government-funded initiative to assist organisations in identifying how they will be affected by climate change so that adaptation strategies can be developed | www.ukcip.org.uk/ |
| Engineering and Physical Sciences Research Council | Development of high-resolution scenarios for the EPSRC/UKCIP Building Knowledge for a Climate Change (BKCC) initiative | www.epsrc.ac.uk |
| Natural Environment Research Council (NERC) | Funding the Rapid Climate Change (RAPID) programme aiming to improve ability to quantify probability and magnitude of future rapid climate change | www.soc.soton.ac.uk/rapid/rapid.php |
| Tyndall Centre for Climate Change Research | National UK centre for trans-disciplinary research on climate change | www.tyndall.ac.uk/ |
| Hadley Centre for Climate Prediction | UK Government centre for research into the science of climate change | www.met-office.gov.uk/research/hadleycentre/index.html |
| Climatic Research Unit (CRU) | The Climatic Research Unit is concerned with the study of historic, present and future natural and anthropogenic climate change | www.cru.uea.ac.uk/ |
| Energy Saving Trust | UK organisation addressing the damaging effects of climate change through the sustainable and efficient use of energy | www.est.org.uk/ |
| Carbon Trust | Action Energy Programme – free, independent, practical advice to business and public-sector organisations on reducing energy use. Funded by UK Government | www.thecarbontrust.co.uk |

There are also numerous UK publications, including UKCIP regional studies, seeking to raise awareness of climate change and likely climate change related effects, including potential impacts on the construction industry. Specific documents for construction professionals have also been published. References to a range of such publications are given in Appendix 5.

## 1.1.4 Awareness of climate change within the UK construction industry

Despite the large amount of information available on the expected scale of impacts from climate change, there is little evidence of mainstream construction companies within the UK implementing, or even considering, possible adaptation techniques or mitigation measures to reduce impacts from climate change. This appears equally true for those maintaining existing infrastructure. The reasons for this are not clear and are likely to be complex, but they may include:

- A low level of awareness of one or several of the following:
  - the likelihood, scale and timing of these changes in the climate
  - the consequential impacts on the built environment
  - the importance of taking decisions now about possible events well in the future
  - the effect on decision-making associated with design for new-build, refurbishment and maintenance, and construction project management.
- The focus of the construction industry is on conformance with published standards. Where standards have been amended in recent years, climate change is likely to have been considered although it is not clear whether this has been addressed to a consistent standard. Furthermore, as construction standards tend to be highly specific, the direct and broader indirect issues of climate change may not have been adequately covered. For example, Part L of the Building Regulations does not discourage the use of air conditioning, which is known to be energy-intensive and consequently has undesirable impacts on climate change.

- The user is not always told why standards have been revised, so opportunities are being lost to raise awareness of the shifting factors affecting the industry, such as climate change.

Developments are needed to enable climate change to be considered appropriately in the construction industry's decision-making processes. A continuing programme of education will help raise awareness and provide necessary information, but wider use of techniques such as risk management will also be important. The awareness of a sample of construction professionals based on a focused survey is discussed in Section 1.3.3.

## 1.2 PURPOSE AND SCOPE

The purpose of this publication is to give practical guidance that will help construction clients (from both the private and the public sector), engineers, contractors and other interest groups to understand the implications of climate change and how to assess and manage the risks associated with it, including ground movement, rain penetration and wind loading, in a similar fashion to procedures already developed for flood-related risks.

---

**POINT TO NOTE**

**Flooding**

The scope of the project was designed to fit in with, and not duplicate, existing projects and guidance. Flood-related issues have therefore been excluded as extensive guidance has been, and continues to be, developed in that area.

---

Parallel research projects to this study include the following DTI Partners in Innovation studies:

- "Mitigating the Effects of Climate Change by Roof Design'" undertaken by BRE, promoting the use of green roofs or roof gardens for combating impacts of climate change, specifically higher summer temperatures and increased winter rainfall (BRE, in press)

- "Climate Change and the Internal Environment of Buildings" undertaken by Arup Research and Development, detailing impacts of climate change to internal conditions, focusing in particular on thermal comfort (Arup R&D, in press).

There is significant scientific information on climate change and its possible effects. Before commissioning this research, CIRIA identified that the majority of this information provides little practical guidance for the construction sector except on flooding issues. Such guidance as does exist has not been set in a risk context.

---

**POINT TO NOTE**

**Risk**

The risks from climate change on the construction industry can be described in the following equation:

RISK OF IMPACT TO CONSTRUCTION PROCESSES = LIKELIHOOD OF CLIMATE CHANGE EVENT (eg increased rain) x SEVERITY OF CONSEQUENCE (eg structural failure of building)

This definition of risk has been applied throughout this book. There is a wide range of issues that should be considered when identifying risks associated with climate change. These include serviceability, maintainability and durability, as well as associated aspects in areas such as quality, environment and health and safety.

---

> **POINT TO NOTE**
>
> **Risk events**
>
> Standard risk management techniques refer to the likelihood of occurrence of individual events. Changes in climate will occur as trends over time rather than as events – for example, increased rainfall or higher temperatures. To maintain integrity between the disciplines of risk management and climate change, this book uses the term "climate change event" to describe the result of trends over time.

## 1.3 METHODOLOGY

### 1.3.1 Phase I: Literature review and industry questionnaire

A literature review was undertaken to establish the climate-related scenarios for the UK and review the effect on the construction industry of a predicted change in climate. This literature review also included a review of risk assessment processes developed for assessing climate change, to investigate their relevance to practical application within the construction industry.

A questionnaire was developed as part of this phase to ascertain (or confirm) the current level of knowledge across construction professionals as well as their perception of the relative importance of climate change to their businesses. This questionnaire is included as Appendix 1.

The questionnaire was disseminated to professionals across several construction sectors. Just over 15 per cent responded. It had been hoped that the responses would point towards those impacts that should be considered in greater detail in Phase II. However, the low response rate meant that the methodology for identifying these issues had to be altered, as described below.

### 1.3.2 Phase II: Identification of specific consequences for the construction industry

In Phase II the specific consequences of climate change for the construction industry were identified and then considered in terms of risk-based decision-making.

The consequences were selected using two qualitative approaches. An initial brainstorming session of the issues identified in the literature review formed the basis of a list of aspects for consideration. These were ranked against predicted climate change events and the probability (or "confidence") of each event occurring. Priority tended to be given to those consequences that would arise from several climate change events, and those to which the UKCIP report had assigned a higher confidence level were generally selected over those that would be affected by events with a lower confidence.

### 1.3.3 Questionnaire results

The questionnaire responses indicated that there was little general awareness of the impacts that climate change could have across the industry, or even which climate change events are predicted to occur. These responses were largely in agreement with previous surveys by the CRISP climate change task group (Lowe, 2001) and the Tyndall study (Hertin et al, 2002). The responses confirmed that the construction industry needed some form of guidance on climate change.

Although the responses from the questionnaires are not considered to be statistically significant, the following observations can be made. Overall, architects appeared more aware of such impacts than other professionals. Those involved in the building sector appeared more aware than colleagues within infrastructure. The questionnaire process may itself have facilitated some awareness. One respondent indicated that the business had not previously taken account of climate change, but would consider this as a parameter in the risk assessment process for future projects. Specific impacts identified by respondents were varied in type and detail and often were largely dependent on the respondent's role and level within an organisation and the construction process with which they were associated.

**Case studies**

Case study examples are used in the book to illustrate points and highlight issues. To minimise duplication of details, this publication provides just an overview but includes a reference to the original document for the interested reader. In some cases, extracts are taken from press reports, not because of their technical accuracy, but to illustrate how the public may perceive events.

## 1.4 READERSHIP

The guidance should be of interest to all those involved in the built environment, but it is specifically targeted at four groups.

1   *Clients and owners* – who need to be assured that their advisers and builders have adequately addressed climate change effects that may affect the performance and life expectancy of their facilities, and that they are comfortable with the attendant risks.

2   *Designers* – professionals concerned with the design of constructed assets of all sorts (from houses to commercial buildings to infrastructure), who need to understand the nature of climate change effects that may arise during the specified lifetime of the asset, and the ways in which risks may be assessed and managed.

3   *Contractors* – who need to recognise that changes to the climate may introduce risks into construction sequences and timings that will affect programming and may, in some circumstances, dictate that work be undertaken in different seasons.

4   *Others* – such as insurers and facility managers, who need to understand how climate change may affect the performance of assets over time and how the attendant risks have been managed.

## 1.5    HOW TO USE THE BOOK

The chapters in the book follow a logical sequence. For the best understanding they should be read in order. However, it is also recognised that readers may have different levels of knowledge and needs, so, for ease of reference, the contents of each chapter are shown in Figure 1.1.

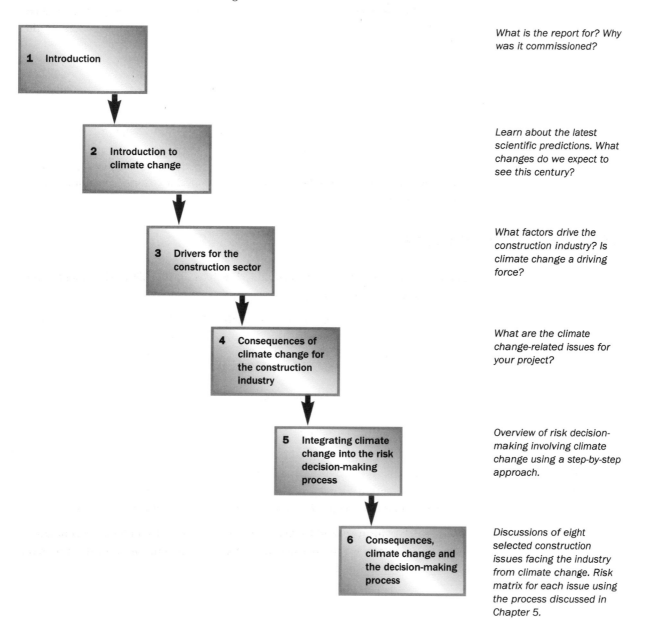

*What is the report for? Why was it commissioned?*

*Learn about the latest scientific predictions. What changes do we expect to see this century?*

*What factors drive the construction industry? Is climate change a driving force?*

*What are the climate change-related issues for your project?*

*Overview of risk decision-making involving climate change using a step-by-step approach.*

*Discussions of eight selected construction issues facing the industry from climate change. Risk matrix for each issue using the process discussed in Chapter 5.*

**Figure 1.1**      *Layout of the book*

# 2 Introduction to climate change

## 2.1 THE CHANGING CLIMATE

It is not easy to perceive changes in climatic conditions, because fluctuations are known to occur within a typical climate pattern on a local (eg UK), regional (eg Northern Hemisphere) and global scale. Nevertheless, scientists have tracked key climate parameters over long periods, which have enabled the overall trends to be viewed. It is these overall trends that are considered in climate change, not annual or monthly variations. For instance:

- global averaged surface level temperatures have risen by about 0.6°C over the past 100 years with about 0.4°C of this warming occurring since the 1970s (Hulme *et al*, 2002)

- the frequency of very hot days has increased since the 1960s (Hulme *et al*, 2002)

- temperature changes experienced in the Northern Hemisphere during the 20th century were greater in magnitude than in any other century during the past 1000 years (Houghton *et al*, 2001).

During 2003, the UK experienced some extreme events that much of the scientific community considers demonstrate its predicted climate change (Box 2.1).

---

**POINT TO NOTE**

**Extreme events**

For the purposes of this publication, the term "extreme" is used to indicate an event, or the limit of a range, that is worthy of particular note. It may, as in the box below, be record-breaking, or it may indicate a quantity or sequence that is quite unusual compared with historical experiences.

---

**Box 2.1** *UK climatic facts 2003*

The year 2003 was the fifth warmest, and included the fourth-warmest summer period, recorded in the UK since reliable records began in 1659. Mean central England temperature was 1.09°C above the long-term average of 9.73°C.

The record for the hottest temperature in England was broken on 10 August when 38.5°C was reached at Brogdale, near Faversham in Kent. In the same month a temperature of 32.9°C broke the Scottish record at Greycrook in the Scottish Borders.

January to October 2003 was the eighth-driest period overall since 1766, but in August, Carlton in Cleveland reported 48 mm of rainfall in just 15 minutes and in November double the average rainfall was recorded in parts of south-east England. Charlwood and Wisley in Surrey recorded more than 75 mm of rainfall in a 48-hour period between 21 and 23 November, which exceeded their average for the whole month.

**Source:** Met Office, 2003

---

The recent occurrences of variations in our climate system that have produced extreme climatic conditions are not limited to the UK. Other countries across the globe have also experienced changing climate patterns such as severe typhoons in Philippines, Taiwan and China, severe hurricanes in the Caribbean and southern USA, winter droughts in Australia and increased flooding and landslides in Bangladesh.

## 2.2    WHY IS OUR CLIMATE CHANGING?

There is general recognition that variations may be expected in natural climate patterns where climate deviates from the "average" known conditions. Through scientific research and evidence built up by analysis of ice cores by organisations such as the British Antarctic Survey, fluctuations are known to have occurred since prehistoric times (Houghton *et al*, 2001). Time-scales over which natural fluctuations can occur vary significantly depending on causation. They may be experienced yearly, over several decades or centuries, while those between an ice age and an interglacial period may occur over thousands of years or more (Houghton *et al*, 2001).

---

**POINT TO NOTE**

**Beyond natural variations**

Evidence suggests that climate is changing beyond its natural variability. Scientists predict that by the end of this century the globally averaged surface air temperature may have risen by up to 5.9°C (Houghton *et al*, 2001, Hulme *et al*, 2002). To put this in perspective, this is the same scale (in reverse) as the temperature change seen during the last ice age.

---

In order to identify the causes of recent changes in climate, the Hadley Centre climate model was used to simulate global climate from 1860 to 2000 considering natural factors, human factors (greenhouse gases and aerosols) and then both sets of factors combined. Only when both sets of factors were combined could the temperature rises in the mid-20th century and, more recently since the 1970s, be explained (Hulme *et al*, 2002).

Human activities are likely to have made a significant contribution in accelerating this rate of climate change particularly since the onset of the industrial revolution (Houghton *et al*, 2001; Hulme *et al*, 2002). The everyday reliance on fossil fuels for the transport, energy generation and similar purposes has contributed to the increase in amount of greenhouse gases released to the atmosphere.

Carbon dioxide (generated by energy production etc), methane (eg from agriculture) and hydrofluorocarbons (notably from refrigeration) are all examples of greenhouse gases. These gases trap energy in the lower atmosphere and, by warming the climate, contribute to changing the pattern of other climatic events including changing rainfall intensity and storm frequency (Houghton *et al*, 2001). Figure 2.1 highlights the change in temperature over the past 1100 years against the 1961–1990 average.

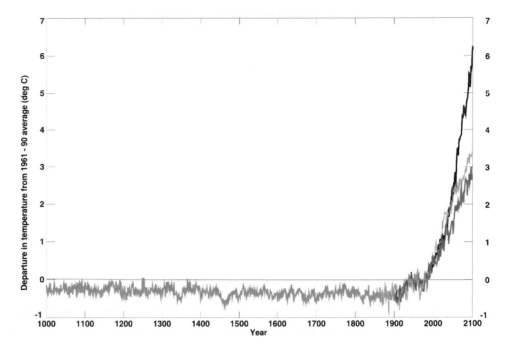

**Figure 2.1**    *Change in temperature over the past 1100 years against the 1961–1990 average*

## 2.3    LIKELY CHANGES TO THE UK CLIMATE OVER THE NEXT CENTURY

To assess and understand likely future changes in climate, computer-based predictive models have been developed that take account of both natural and man-made factors. Designing and developing these models is highly complex, as any alteration in man's activity is likely to affect the outcome (see Box 2.2).

**Box 2.2**    *Examples of challenges facing climate change scientists*

> The climate system reacts over a long period. Considerable attention has been focused at the global level (Kyoto Protocol) and within the UK (Climate Change Levy) to reduce the amount of greenhouse gases produced. Despite these best efforts our climate is likely to continue altering for many decades. Scientists calculate that the climate change we are seeing now results from the industrial activities of around 100 years ago. It is generally accepted that predictions for the next 30–40 years are relatively certain, as these are based on greenhouse gas emissions already released. Beyond this date the predictions can vary widely, depending on the degree to which we change our practices to reduce emissions of greenhouse gases.

To address uncertainties in emissions of greenhouse gases the IPCC has developed a series of scenarios that reflect four alternative routes of world development (McCarthy *et al*, 2001). These take account of the likely amount of greenhouse gas emissions for each route, which assisted UKCIP in designing the details and the specific input parameters that were used in the UK models. As well as the uncertainties about future emissions, climate change modellers are also exploring more accurate ways of modelling the many complex processes that control our climate.

The outputs from the models provide a range of estimates for various components of the climate, such as temperature and rainfall, and are given for the 2020s, 2050s and 2080s (Hulme *et al*, 2002). Levels of confidence have also been assigned to particular qualitative statements in terms of high, medium or low. These confidence levels reflect the reliability of a selection of the predicted outputs (scenarios) based on the subjective opinion of the developers of the models and are not prescribed to be scientifically accurate (Hulme *et al*, 2002).

**POINT TO NOTE**

**Model outputs**

Outputs from the models are presented:

- in a series of 30-year periods:
  - 2020s = the period from 2011 to 2040
  - 2050s = the period from 2041 to 2070
  - 2080s = the period from 2071 to 2100
- in terms of seasonal averages and also extreme events. It is anticipated that the extreme events may have the biggest impacts on construction projects.

Regional variations within the UK are included and presented as maps within the UKCIP report, examples of which are shown as Figures A2.1–A2.4 in Appendix 2. Figures A2.1 and A2.2 highlight the changes in average seasonal temperatures and annual temperatures for the 2020s, 2050s and 2080s for the high and low emissions scenarios respectively. Figures A2.3 and A2.4 highlight the changes in average seasonal and annual precipitation for the 2020s, 2050s and 2080s for the high and low emissions scenarios respectively. These figures clearly illustrate the regional variation across the UK.

Many variables exist in modelling future climate change and it is difficult to know which scenario is more likely to occur or which event will have the greatest impact. This report does not suggest or indicate the likelihood of one particular scenario occurring and as such does not provide in-depth discussions of each predicted scenario. What it does is to argue that the uncertainties need to be incorporated into overall risk management processes that are becoming a regular part of construction projects and to indicate how this should be done.

A summary of some of the predicted events is provided in Table 2.1. For further information the reader should consult the source document of climate change scenarios published by UKCIP.*

**POINT TO NOTE**

**Warming or cooling?**

The majority of the scientific community involved in climate change predicts a rise in temperatures in the UK and this opinion has formed the basis for the study and is reflected in this publication. It is recognised, however, that an alternative school of thought within the scientific community considers that the effect of climate change may be a considerable cooling of the UK climate.

The Gulf Stream (or North Atlantic Conveyor) is the ocean current responsible for keeping the British Isles much warmer than landmasses in equivalent latitudes (Labrador, for example). It is part of a circulation of ocean water that transfers heat from the Equator to the North Atlantic. It has been suggested that global warming could "switch off" this ocean current, which would lead to cooling of the UK climate. Although there is good evidence that the Gulf Stream has weakened considerably in the past, it has remained stable for about the last 8000 years. The probability that global warming could lead to a complete halt of this ocean current is very small but cannot be completely disregarded. Nevertheless, the current consensus is that there is little likelihood that the Gulf Stream will shut down and lead to significant cooling of Europe, at least over the next 100 years.

**Source:** UKCIP, pers comm.

---

\* The UKCIP Report 2002 (Hulme *et al*, 2002) was developed by the UK Climate Impacts Programme, Hadley Centre for Climate Prediction and Research and the Tyndall Centre for Climate Change research.

**Table 2.1**  *Summary of the UK's likely climate change scenarios. Source: Hulme et al, 2002.*
*(↑ = increase; ↓ = decrease; blank = information not available)*

| FACTOR | PARAMETER | CHANGE | | | SUMMARY | CONFIDENCE |
|---|---|---|---|---|---|---|
| | | 2020s | 2050s | 2080s | | |
| Temperature | Annual mean temperatures | 0.5–1.5°C | 1–3°C | 1.5–4.5°C | Annual warming by 2080s by 1–5°C depending on region and scenario | High |
| | Summer mean temperatures | 0.5–1.5°C | 1–3.5°C | 1.5–5°C | SE likely to experience greater warming than NW, particularly in summer | High |
| | Winter mean temperatures | 0.5–1°C | 1–2°C | 1.5–3.5°C | Greater night-time than daytime warming in winter | Low |
| | Extreme summer temperatures | | | Approx 10-fold ↑ in no of very hot days | This is presented for the medium-high emission scenario only | High |
| Precipitation | Annual precipitation | 0–10% ↓ | 0–10% ↓ | 0–20% ↓ | | |
| | Average summer precipitation | 0–20% ↓ | 0–40% ↓ | 0–60% ↓ | Much drier summers for whole of UK | Medium |
| | Average winter precipitation | 0–15% ↑ | 0–25% ↑ | 0–40% ↑ | Generally wetter winters for whole of UK | High |
| | Extreme winter precipitation | | | 100% ↑ | This is presented for the medium-high emission scenario only | High |
| Seasonality | Annual | | | | More Mediterranean-like climate regime | |
| | Precipitation | | | | Greater contrasts between drier summers and wetter winters | High |
| Cloud cover | Annual | | | 0–9% ↓ | | |
| | Summer | | | 3% ↑ to 18% ↓ | Reduction in summer and autumn cloud especially in S, and increase in radiation | Low |
| | Winter | | | 3% ↓ to 6% ↑ | Small increase in winter cloud cover | Low |
| Relative humidity | Annual | | | 0–9% ↓ | | |
| | Summer | | | 0–18% ↓ | Relative humidity decreases in summer | Medium |
| | Winter | | | 0–3% ↓ | | |
| Average wind speed | Annual | | | 0–7% ↑ | | Low |
| | Summer | | | 11% ↓ to 13% ↑ | | Low |
| | Winter | | | 3% ↓ to 13% ↑ | | Low |
| | Occurrence of storms | | | Winter depressions may ↑ from 5 to 8 | This is presented for the medium high emission scenario only | Low |
| Snowfall | Winter | | | 30–100% ↓ | Totals decrease significantly everywhere | High |
| Soil moisture | Annual | | | 0–20% ↓ | | |
| | Summer | | | 0–50% ↓ | Decrease in summer and autumn in SE | High |
| | Winter | | | 0–8% ↑ | Increases in winter and spring in NE | Medium |
| Sea level rise | Global average sea level change | 4–14 cm | 7–36 cm | 9–69 cm | Change in the average level of the sea relative to the land will not be the same everywhere because of natural land movements and regional variations in the rate of climate-induced sea-level rise. NB This information is the IPCC range, which represents global information | |
| | Change in return period for extreme high water levels | | | Return period for extreme high water levels will shorten; eg for Immingham, 1 in 20-year event becomes a 1 in 7-year event in 2080s | | |

# 3    Drivers for the construction sector

From the information reviewed in Chapter 2 it is clear that some aspects of climate change are already being seen and have demonstrable impacts on the built environment, eg flooding. It is therefore realistic to assume that other climate changes will have an impact on the construction industry in the next 20–50 years, if not sooner, and certainly well within the design lives of many facilities.

The construction industry in the UK is primarily driven by government, guidance and standards (such as British Standards and Building Regulations), client specification and contracts, and the insurance industry. These drivers are all likely to be responsive to climate change and so will affect the construction industry.

---

**POINT TO NOTE**

**Health and safety**

This publication is principally concerned with physical changes that will affect the built environment. There are nevertheless indirect effects that may influence decision-making in construction. For instance, higher average summer temperatures, more very hot days and higher UV concentration in sunlight will have implications for worker health and efficiency, through long-term effects such as increased skin cancer and shorter-term consequences in a slower work rate. Both represent risks that could have financial implications. Learning will be required from countries that operate under these conditions at present.

---

## 3.1    GOVERNMENT AWARENESS AND CHANGING GUIDANCE

Few construction guidance documents either in the UK or more broadly across Europe clearly indicate that climate change has been considered in their development. Some recently been amended Building Standards have taken climate change into consideration, however, including Approved Documents A (structure) and C (contamination and moisture) of the Building Regulations in England. The new Eurocodes do not specifically discuss climate change, but there is provision for regional variations in climate conditions, such as wind or snow maps, to be taken into consideration in their application (Eurocodes, 2003).

Given the time and money the UK Government is spending on research into the implications of climate change, it is likely that climate change will continue to feature in the development and amendment of guidance and standards. Box 3.1 shows how severe flooding events in the UK drove the development of related guidance.

**Box 3.1**    *Case study: planning guidance – embracing climate change*

---

In April 1998 severe flooding occurred in the Nene and Great Ouse valleys. In Northampton more than 10 000 people were directly affected. It was thought that this event might have been caused or exacerbated by climate change, and there was a review of Planning Policy Guidance, with the publication of PPG 25 in July 2001 (DTLR, 2001). Even before it was published, further widespread flooding occurred in autumn 2000. For the first time in a PPG, guidance is given on the possible impact climate change may have on the scale and likelihood of flooding in the UK. This PPG has ensured that flooding is now actively considered in all planning decisions.

---

## 3.2 CLIENTS AND CONTRACTS

The past 20 years have seen the development of several new forms of contract to address clients' needs for more effective ways of procuring new infrastructure. For example, the Private Finance Initiative has encouraged private investment in public projects. Integral to the development of these new contracts is the need to balance risk fairly, so that clients can achieve reasonable certainty of cost and contractors are able to work at a reasonable profit.

The risks of impacts from climate change will assume increasing importance in the pricing of such projects, and it will be vital for all parties to the project to understand how to assess and manage those risks. These could affect not only the quality, durability and cost of the finished product, but also the way in which it is built.

> **POINT TO NOTE**
>
> **Whole-life costing**
>
> The principles of whole-life costing (WLC) are increasingly being used to assess major construction projects. The consequences of climate change on material durability and the cost of maintenance will be another factor to be introduced into WLC models when developing the cost benefits of various options for such projects.

## 3.3 INSURANCE INDUSTRY

In the UK between 1990 and 2000, weather-related insurance claims totalled between £360 million and £2.1 billion a year (ABI, 2003). With the high current value of housing and evidence of an increase in frequency and intensity of extreme climate events, the impact on property will affect both insurance providers and the insured. These factors in combination have already increased the cost of claims. Furthermore, homes in places where risks are unacceptably high and no defence improvements are planned may no longer be guaranteed insurance. The government and the Association of British Insurers (ABI) had guaranteed insurance cover to those who live in homes on floodplains for two years after the floods of autumn 2000, but this agreement ended at the start of 2003 (UK Home Insurance, 2003). It has been superseded by the ABI "Statement of Principles", which aims to reassure homeowners in floodplains about the continued availability of insurance.

The National House Building Council (NHBC) provides insurance for 10 years against structural damage including subsidence and heave for new and renovated buildings where works have been undertaken by registered builders (NHBC, 2003a). In 1999, the ABI instigated research into subsidence damage and associated insurance approaches in other countries. The report found that:

- insurance is not always available for subsidence in countries other than the UK (ABI, 1999)
- UK homeowners have a lower tolerance to minor subsidence-related damage than do homeowners in other countries (ABI, 1999). Box 3.2 identifies the cost of subsidence in the UK and experience in other countries.

It may not be unreasonable to assume that, if property damage from climate change-related events continues, property insurance cover will be limited for other risks as it has been recently for flood risk. Should this happen, property-owners would wish to have increased confidence that the building design was appropriate.

**Box 3.2**     *Case study: financial implications of subsidence and heave*

The insurance industry in the UK currently absorbs the cost of subsidence and heave-related damage. Subsidence in the UK costs the insurance industry an estimated £400 million each year (Graves and Phillipson, 2000). For each dwelling affected by subsidence and heave this can mean a cost of approximately £5000–50 000 to remedy.

In other countries the cost of subsidence and heave repairs borne by the insurance industry differs markedly from the UK. In South Africa insurance companies are progressively removing subsidence and heave cover from insurance through interpretations of existing policies and by adding exclusions to new policies. Subsidence and heave is also a large problem in Australia, but house insurance policies always preclude subsidence and heave cover (cited at <http://www.willowmead.co.uk/news/abi.htm> from ABI, 1999).

Readers should note that designers and contractors may not be relieved of responsibilities relating to subsidence and heave damage. Experience from outside the UK indicates that in some cases contractors are liable for subsidence and heave damage in new houses and they can be forced to rectify subsidence and heave-related damage if they are found not to have used the most appropriate foundations and techniques to minimise post-construction damage (Supreme Court of Queensland, 1992).

## 3.4     USER AND PUBLIC PERCEPTION

Building stock and infrastructure currently being developed typically has a design life of more than 60 years and so should have the capacity to withstand predicted climatic events over its entire design life. Users of infrastructure and buildings already expect that best practice will have been implemented at the time of design (or redevelopment/ refurbishment). It is surely not unreasonable for a user or the public at large to assume that "best practice" will have taken into account predicted changes in conditions to the climate over the facility's designed lifetime. If a structure is unable to withstand expected climate conditions within its design life, one may question whether it can be considered "fit for purpose".

# 4 Consequences of climate change for the construction industry

A multitude of potential impacts or consequences for the construction industry could arise from the climatic events shown in Box 4.1. Some, such as flooding, are already being witnessed.

**Box 4.1**     *Predicted changes in our climate*

The following list identifies the major average climatic changes and extreme events the construction industry should be aware of (for confidence levels and estimated changes refer to Table 2.1):

Averages:

- increase in average summer and winter temperatures
- decrease in average summer rainfall
- greater seasonality
- decrease in cloud cover and hence an increase in solar radiation
- decrease in relative humidity
- slight increase in wind speed (particularly across southern England)
- decrease in snowfall
- decrease in soil moisture (particularly during the summer)
- increase in sea levels around the UK coastline.

Extremes:

- extreme summer temperatures
- extreme winter precipitation (volume and intensity)
- increase in occurrence of winter storms (rain and wind combined)
- extreme high sea levels will become more frequent.

Diagram 4.1 below identifies some of the key climate change issues and the possible consequences for the construction sector.

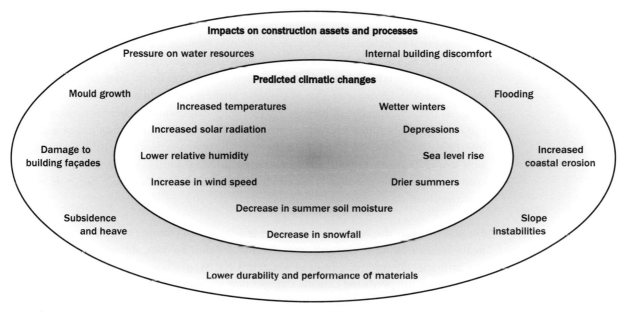

**Figure 4.1**     *Predicted changes in climate and associated impacts for the construction industry*

The following sections summarise some of the consequences of climate change for the construction industry. This list is not intended to be exhaustive, but is indicative of issues that need to be taken into account at design stage to ensure the integrity of the facility, and that will affect construction processes by comparison with the way they are undertaken today. There are other issues, such as the effects on sealants or plastics products, where guidance will have to be sought from manufacturers. Some of the following items are discussed in more detail in Chapter 6 to demonstrate how the proposed risk management process can be applied.

## 4.1 DELAYS TO THE CONSTRUCTION PROCESS

The construction industry is largely based around site work and as such is vulnerable to climatic conditions. Key predicted climate change events for the construction process include heavier winter rainfall, rises in temperature and an increase in solar radiation. Extreme climatic events have been known to disrupt and even halt site-based activities, leading to longer construction times and associated increases in costs, damage to materials and ultimately late delivery (Graves and Phillipson, 2000). Climate has been identified as the fourth most likely reason for delays to highway construction in the USA (Ellis and Thomas, 2002).

Effects on construction processes resulting from climatic events could be relevant to projects of all scales and types. Theoretically, interruptions and delays could be limited by planning construction activities for seasons with the most suitable climatic conditions. For larger projects it is unlikely that this will be completely viable option, however. It may be possible to schedule smaller projects to avoid the least favourable weather conditions completely, although, conversely, if appropriate scheduling or rescheduling is not possible, significant impacts may occur. Therefore in terms of relative impacts, small-scale projects may be affected to a far greater extent than large projects. For example, a storm encountered for five days may result in a small project being completely behind schedule, with consequent effects on budget, while causing only minor disruption to a major, year-long project with a large budget.

## 4.2 SUBSIDENCE AND HEAVE

No one wishes to experience structural damage to their home. One aspect of climate change that predominantly affects housing, and hence the general public, is subsidence (and its corollary heave, which can produce similar consequences). Subsidence can cause internal or external cracking, and structural damage may occur where clay soils are both predominant and particularly reactive to alternate dry and wet periods. Evidence from NHBC (personal correspondence) indicates that a long dry period (two dry summers and a dry winter) may pose the greatest risk. While the UKCIP data suggests that climate change is unlikely to produce drier winters, there is concern that the more intense rainfall anticipated may cause flash flooding, resulting in less water infiltrating the ground. In such a situation we might expect less re-saturation of soils over winter. Subsidence and heave is already a problem within the UK and can be costly to repair. The predicted climatic changes are likely to exacerbate the problem, particularly in the south-east of England where clay soils predominate.

For several years the NHBC has provided guidance and comment on subsidence and heave issues relating to clay and trees (NHBC, 2003b). To date, subsidence and heave has primarily affected properties in which this NHBC guidance has not bee taken into account, but the predicted climatic changes may increase the incidence.

**Box 4.2**     *Case study: subsidence on roads – adaptation to climate change*

> Subsidence problems on roads across eastern England after the long, hot summer of 2003 is likely to have raised awareness of the implications of climate change on transport infrastructure at local government level. Rather than motorways and A-roads, rural roads suffered the biggest impact of subsidence, as less resistant pavement materials could not withstand the dry summer and the subsequent shrinkage of the underlying clay soils. The cost to Cambridgeshire Country Council alone is predicted to exceed £1 million. Councils are looking at adaptation techniques to ensure road pavement materials are robust enough to withstand the changing climate of drier spells in summer followed by intensive rainfall in the winter.
>
> **Source:** *NFU Countryside*, Rural News, Monday 10 November 2003.

## 4.3     DECREASED DURABILITY AND PERFORMANCE OF MATERIALS

Materials used in construction are designed to last either for as long as the building itself (based on its predicted lifespan) or for a calculated period that keeps maintenance costs at a minimum over the building's lifespan. Understanding of the inter-relation between materials and climate is important in ensuring that any potential impacts are minimised. Predicted climate change events of most relevance include: increased driving rain, possible increased strong wind episodes, increases in temperature, increased solar radiation and decreases in relative humidity.

If a material on an existing building or infrastructure item fails or is damaged before the end of its expected lifespan, the associated maintenance cost will rise. An example would be the rapid degradation of a window seal caused by increased solar radiation. Increased maintenance frequency, or early replacement of the material because its life has ended prematurely, will increase the costs associated with maintenance of buildings and also impact on the effective lifespan of buildings and infrastructure.

## 4.4     SLOPE INSTABILITIES

Instability within slopes, both natural and man-made, is a global problem. It has caused extensive damage to transportation infrastructure, commercial and domestic structures and, in the most extreme cases, has resulted in loss of life.

Slope instabilities, and the resulting movement of the landform, are a common problem in the United Kingdom, and they occasionally result in catastrophic and well-documented failures.

The presence of water within natural slopes and man-made slopes without drainage is a primary cause of slope instability. Intense or prolonged periods of rainfall may markedly increase moisture within slopes. The predicted increases both in frequency and intensity of winter rainfall fall are likely to increase significantly the occurrence of slope instability across the UK. Those areas historically considered as unstable will be at even greater risk of failure. As noted in Box 4.3, this could have a significant impact on the existing built environment, causing extensive damage to residential dwellings, commercial properties and the transportation infrastructure.

The major infrastructure owners are alive to the dangers of slope instability, are vigilant in monitoring and actively employ preventative techniques. The greatest problems are likely to affect those owners who are less aware of the risks inherent in slope instability.

Although the predictions of a build-up of water pressures in slopes appear to conflict with comments in Section 4.7 about the lack of infiltration that also has been predicted, in fact much depends on the underlying geology. Slope instabilities occur most widely in clay soils where permeabilities are low and small volumes of water can give rise to high pressures. In contrast, groundwater resources are held in permeable soils and rocks, such as limestones, sandstones and chalk, which tend not to be unstable in slopes. Replenishment of groundwater resources requires the infiltration of large volumes of water.

**Box 4.3** *Case study: challenges to managing slope instability of UK's railway infrastructure*

The changing climate could cause embankment failure. Most of the UK's railway infrastructure was built in the 19th century, when embankments were constructed steeper than they would be today. These are precisely the sorts of structures that are susceptible to more frequent landslides than the equivalent modern ones. Not only will some of the embankments have to be reprofiled, but designers will also have to consider the timing of such works. Prolonged wet weather and intensive rainfall during earthworks can cause problems if absorbent materials are used, as these can retain water and become weak, limiting their usable lifetime. Delays would increase costs and extend the timing of such projects. If, to minimise delays, less absorbent materials were used, the ultimate cost of the solution might be increased.

**Source:** Beazant, 2003

## 4.5    DAMAGE TO THE FABRIC OF BUILDINGS

The external fabric of a building is the first line of defence against the climate. Driving rain, increased winds or more strong wind episodes and an increase in solar radiation all affect the weathering rates and durability of building façades.

This area of risk will have a major impact on new-build properties, and, potentially more importantly, on those parts of Britain's existing building stock with significant residual lives, necessitating substantial additional maintenance. Building designs generally evolve in sympathy with climatic conditions experienced within regions. To compensate for intense rainfall for example, Scottish building design has evolved to include recessed window detailing, rendering and overhanging eaves (Graves and Phillipson, 2000). Predicted changes in climate, however, may not have been accounted for in existing designs in many regions. This regional evolution can also provide useful information, for we may be able to learn from successful adaptations in other countries.

Damage to, or increased weathering of, façades can also alter the internal conditions of those buildings. Driving rain that penetrates the exterior of a building may increase the potential for damp, bringing with it the risk of associated mould problems, which in turn may injure the health of the building occupiers. There is also the risk that increased condensation and water penetration into the fabric of the building may accelerate the corrosion of vulnerable metallic items, such as brackets, fixings and frames, unless these have been adequately designed or provided with protective coatings.

Another DTI Partners in Innovation project, "Mitigating the effects of climate change by roof design", promotes the use of the green or garden roof. It is suggested that green roofs are better able to deal with the predicted higher summer temperatures and to manage increased rainfall in winter by buffering runoff during storms, reducing the load into the drainage system. The research is being carried out by BRE and will be published in 2005 (BRE, in press).

## 4.6      STRUCTURAL DAMAGE FROM WIND-RELATED EVENTS

The UK is one of the windiest countries in Europe and every year many structures and buildings are damaged by wind-related events (Graves and Phillipson, 2000). The average number of dwellings damaged each year is approximately 200 000 (ABI, 2003). Wind damage varies widely, from tiles being dislodged from roofs of domestic dwellings to significant damage of large structures.

The increase in wind speed associated with climate change is predicted to be minor in the UK and is an area of low confidence, but the potential costs associated with failure can be significant. There may also be increased storm-related events caused by more frequent depressions (lows), though the confidence in the likely occurrence of this is low.

**Box 4.4**      *Case study: windstorm damage to domestic property*

> The ABI has identified damage to domestic property from windstorms as the most destructive of climate extremes. Considerable losses occur every year in the UK and in major windstorm events the damage can be substantial. The storms of October 1987 and January/February 1990 alone resulted in insured losses of £1.4 billion and £2.1 billion respectively.
>
> The exteriors of a high percentage of UK homes have been identified as being in disrepair and are more susceptible to wind damage, along with older buildings, particularly those built before 1944 when no wind loading standards were in place.
>
> British Standards for wind loading have doubled since inception for some types of building, and tripled within some regions, particularly in the south-east and eastern regions of the UK. Design codes for south-east England may need to increase wind loading requirements by 10 per cent, based on the information provided in the UKCIP02 climate change scenarios.
>
> **Source:** ABI, 2003.

## 4.7      PRESSURE ON WATER RESOURCES

The UK is currently fortunate to experience enough rainfall, dispersed throughout the year, to have access to adequate sources of water supply. Predicted climatic scenarios could change the availability of water resources, particularly in the summer months, when temperatures are likely to increase water demand and extensive drier periods become more common. Although average annual rainfall in the UK is not predicted to fall by more than 10 per cent by the 2050s (Hulme *et al*, 2002), the ability of the natural environment to capture the same amount of water is uncertain. Summer rainfall is predicted to decline and, if predicted increases in rain intensity in winter occur, more rain may run off directly to watercourses, leaving less to infiltrate the ground. In areas reliant on groundwater resources, less water may be available for use. Many processes associated with the construction industry are heavily reliant on water, so any limitation on the availability of water is likely to have an effect both on the construction of buildings and infrastructure and on its use within existing premises. Water resource issues may also start to determine the location of development because of the extra cost of providing water to areas of shortage.

**Box 4.5**      *Case study: pressure on water resources*

> In 2003, a contractor sought approval from the Environment Agency to abstract water from a local river for use in the construction of a road. For a while, this was permitted. But, with little rain, the river levels dropped and permission was withdrawn, and the contractor had to make other arrangements. However, the same dry period led to benefits overall, as earth moving was able to progress earlier in the year than had been planned.

## 4.8 POORER CONDITIONS IN THE INTERNAL ENVIRONMENT

Health and comfort issues associated with poor internal conditions in buildings or dwellings are of primary concern for the construction industry. Internal building environments are largely controlled through design, but they are directly affected and influenced by external climatic factors. Existing buildings may not have been designed to cope with predicted climate change events.

The impacts of poor conditions in the internal environment are many and varied and often hard to quantify. Building occupants can suffer a range of ill-effects, from simple discomfort from rising temperatures, higher wind speed and subsequent draughts, to heat stress and dehydration in hotter and drier periods (Graves and Phillipson, 2000). Other, well-documented indirect impacts may include an increase in biological hazards (Graves and Phillipson, 2000), such as the proliferation of pests in changed ecological conditions, increasing incidence of damp problems and consequent mould growth.

The results of another Partners in Innovation project, "Climate change and the internal environment of buildings", which is being led by Arup Research & Development, will be published in 2005 as a CIBSE technical memorandum (Arup R&D, in press).

## 4.9 FLOODING

Climate change has important implications for the assessment of flood risk and the design of mitigation measures such as flood defences. It is predicted that winters in the UK will become wetter, which could increase the chances of flooding during this season. Rises in sea levels and changes in the frequency and severity of storms are also likely to exacerbate the risk of flooding. Predicted changes in the UK climate mean that flood defences designed in the past to provide protection against a specified level of flood probability will no longer meet the design specifications. Price and McInally (2001) estimated that coastal defence levels in Scotland designed to the 1 per cent annual probability (1 in 100 chance) standard in 1990 would provide a standard of protection of only 5–10 per cent annual probability (1 in 10 to 1 in 20 chance) by the 2050s. For fluvial flood defence standards, they estimated that the 1990 1 per cent flood defence standard would be equivalent to approximately 1.5 per cent (1 in 60 to 1 in 65 chance) by the 2050s and around 2 per cent (1 in 40 to 1 in 60 chance) by the 2080s.

Because of the wealth of existing information on flooding and the effects of climate change on flooding, this publication has omitted consideration of flooding. For further information readers are referred to the sources identified in Box 4.6 below and in A5.

**Box 4.6**      *Information on flooding*

Defra has overall policy responsibility for flood and coastal erosion risk in England. The Environment Agency is responsible for inland and coastal flood defence. Together, Defra and EA sponsor strategic research into flood management as part of a Joint Flood and Coastal Defence Research and Development Programme, which contains six key themes of research: Process, Policy, Broad-scale modelling, Flood forecasting warning, Risk and Engineering.

Information on the Defra/EA joint flood and coastal defence R&D programme can be found on the Defra website, <www.defra.gov.uk>, and EA's website, <www.environment-agency.gov.uk>.

The Foresight flood and coastal defence project is a major research programme that analyses risks of flooding and coastal erosion for the UK between 2030 and 2100. The project's aim is: "...to produce a long-term vision for the future of flood and coastal defence which takes account of the many uncertainties, but which is nevertheless robust, and which can be used as a basis to inform policy". It is being sponsored by the Office of Science and Technology.

For CIRIA's research and guidance on flooding go to <www.ciria.org/theme.htm?ThemeIDNo=6>.

# 5     Integrating climate change into the risk decision-making process

## 5.1    INTRODUCTION

Successful management of risk is relevant to all aspects of construction. The risks for the construction sector as a result of climate change have to be considered along with all other project risks.

The majority of this chapter looks at developing for the construction industry a risk assessment process that incorporates climate change considerations. It is important that risk assessment is considered as only one part of a risk management framework or decision-making process. Together with the Environment Agency, UKCIP has recently developed a framework specifically to help manage climate change risks. The UKCIP decision-making process includes eight stages that could be applied to any risk management process. These are shown in Figure 5.1, which is taken, with permission, from the UKCIP report (Willows and Connell, 2003).

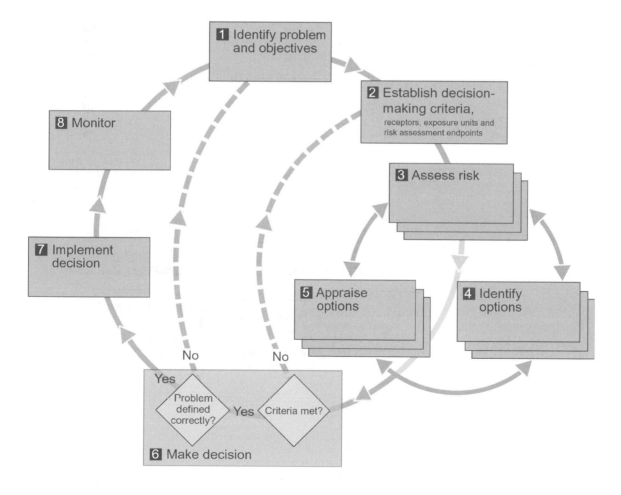

**Figure 5.1**    *A framework to support good decision-making in climate change*

## 5.2     DEFINE THE NATURE OF THE DECISION PROBLEM

Before assessing the risks, both the problem and the desired outcomes or objectives need to be identified. The climate change risk decision-making process and objectives should be aligned with stakeholders' overall attitude to risk. Stakeholders may include the client, insurers, users and the construction supply chain. Risk-based decision-making forms a fundamental part of any decision in the construction industry. Climate change is thus only one component in the decision-making process, although it is one that will demand ever-greater attention in the future. Box 5.1 provides some specific reasons for considering climate change as part of a broader risk-based decision process.

**Box 5.1**    *Why should a risk management process include climate change?*

- To consider what might happen in the future; eg will there be an increase in the risk of incidence of subsidence?
- To consider the consequences if an event occurs; eg what is the changed risk of cosmetic and structural damage to homes?
- To ascertain the possibility of an extreme event occurring. An increase in average temperatures is very likely – in 2003, the UK experienced the hottest day on record.
- To provide a common approach for comparing a wide variety of options
- To manage and reduce likely risks; eg for new buildings, use foundations that will mitigate the risk of movement in the soil and so reduce the likelihood of subsidence.
- To record that the risks associated with climate change have been assessed.

## 5.3     RISK ASSESSMENT

Risk can be defined as a measure of the significance of uncertain or unpredictable events as perceived by a potentially affected party. The significance of the risk can be assessed by combining the likelihood of the event with the severity of the outcome of that event. In terms of climate change the following equation can be applied:

$$\text{RISK OF IMPACT TO CONSTRUCTION} = \underset{\text{eg increased rain}}{\text{LIKELIHOOD OF CLIMATE CHANGE EVENT}} \times \underset{\text{eg penetration of rain, dampness}}{\text{SEVERITY OF CONSEQUENCE OF EVENT}}$$

Based on this definition, to assess the level of risk associated with climate change events the following information is needed.

1  Relevant climate change events and their likelihood.

2  Potentially affected activities during the whole construction process and/or in use.

3  Associated consequences to these activities resulting from climate change events.

As part of this study, an example risk matrix has been developed to help facilitate inclusion of climate change factors in the risk decision-making process. The following sections describe the steps required to enable completion of this matrix; a blank form of the matrix is provided in Appendix 3. Organisations that have existing tools for risk decision-making should align the issues in the matrix with their existing processes. The UKCIP report provides further information on each step of a more detailed risk decision-making process.

**Table 5.1**    *Example of risk matrix table (Appendix 3)*

| Project: | | | | | | | | |
| Location: | | | | | | | | |
| Relevant climate change event | Confidence of prediction of change | Likelihood of climate change event | Construction process | Possible consequences of climate change | Significance of consequence on construction industry | Overall risk of impact | Possible adaptation strategy | Residual risk following mitigation |
| | | | | | | | | |
| | | | | | | | | |

## 5.3.1    Relevant climate change events

In Chapter 2, several predicted climate change events were identified and discussed. Table 2.1 provides a useful summary of the predicted climate change events and the modelled annual average changes and extreme events for the periods 2020s, 2050s and 2080s. The key predicted change climate issues noted include:

- temperature
- precipitation
- seasonality
- cloud cover
- relative humidity
- average wind speed
- snowfall
- soil moisture
- sea level rise.

It is important in any risk assessment process to consider the confidence level associated with predicted changes. For climate change, the scientific community has identified the confidence that it applies to various changes, most particularly associated with the broad qualitative statements rather than detailed quantitative data. For instance, there is a high level of confidence associated with increased annual average temperatures, and a low confidence associated with most of the qualitative statements relating to changes in wind speeds. Table 2.1 provides the level of confidence assigned by the scientists for the climate change events identified above. The predicted climate change event and its associated confidence are two of the elements that form inputs to the suggested risk matrix.

## 5.3.2    The likelihood of climate change events

Having identified the relevant climate change events to be considered, it is necessary to identify the likelihood of these events occurring. At present, climate change modellers have modelled the likelihood based on only one of the four scenarios discussed in Section 2.3, which is the medium-high emissions scenario, and the outputs from this are presented in Table 5.2. The table is a combination of the percentage of years likely to experience specific climate change events (2020s, 2050s and 2080s) and the probability that an event will occur (2080s only). It is important to note that further developments are planned and reference should be made to the UKCIP website (<http://www.ukcip.org.uk>) to confirm the availability of such information in the future. The BETWIXT project, which is funded by EPSRC in collaboration with UKCIP, will provide further information about changes in extreme climate conditions, and can be found at <http://www.cru.uea.ac.uk/cru/projects/betwixt/>.

**Table 5.2**    *Percentage of years experiencing extreme seasonal anomalies across the UK, based on the medium-high emissions scenario (Hulme et al, 2002)*

| Extreme event* | Likelihood of years experiencing event (%) | | |
|---|---|---|---|
| | 2020s | 2050s | 2080s |
| A hot August (average temperature > 3.4°C above average) | 1 | 20 | 63 |
| A warm year (average temperature > 1.2°C above average) | 28 | 73 | 100 |
| Drier summers (rainfall < 37% of average) | 10 | 29 | 50 |
| Wetter winters (rainfall > 66% of average) | 1 | 3 | 7 |
| Increase in number of depressions per season in winter (baseline of 5 per winter season) | ♦ | ♦ | 63 |

**Notes**

\*    Compared with 1961–1990 averages

♦    Information not yet modelled by UKCIP.

Using the values in Table 5.2, it will be possible to make an estimate of the likelihood of the climate change event and populate the risk matrix column entitled "Likelihood of climate change event".

When considering the likelihood of a climate change event it is important to relate the construction process to the correct timescale. Throughout this book, likelihood figures for the 2020s have been used for calculation of risk for construction processes and figures relating to the 2080s have been used for calculation of risk for design processes. This is exemplified by the hotter summer days scenarios where, for design, the likelihood of climate change would be 63 per cent whereas for construction processes the likelihood would be 1 per cent, reflecting the difference between the 2080s and 2020s figures.

### 5.3.3  Potentially affected construction activities

Throughout the whole construction process – ie design, construction and facilities maintenance – there is a wide variety of activities that could be affected by changes in our climate. It is important to identify as many of these as possible in risk-based decision-making to enable high-risk activities to be considered appropriately. Table 5.3 highlights various construction activities or elements that could be affected by climate change, which will assist in the identification of potential consequences or impacts.

**Table 5.3**     *Construction activities that could be affected by climate change*

| | | Construction element or process | | |
| :--- | :--- | :--- | :--- | :--- |
| | | *Design* | *Construction* | *Refurbishment/ facilities maintenance* |
| **Possible climate event** | *Hotter summers* | • insulation<br>• passive cooling<br>• natural ventilation/ active cooling<br>• window glazing<br>• use and selection of materials<br>• foundation design | • durability of materials<br>• external working<br>• placing and curing of concrete | • window glazing<br>• passive cooling<br>• natural ventilation/ active cooling |
| | *Drier summers* | • structural design<br>• use of water<br>• foundation design | • use of water<br>• external working | • use of water<br>• external working |
| | *Wetter winters* | • stormwater drainage<br>• roof drainage<br>• cavity fill insulation<br>• foundation design<br>• structural design<br>• slope design | • external working<br>• location on a flood-plain<br>• slope design<br>• use and selection of materials | • structural design<br>• slope design<br>• use and selection of materials |
| | *Higher sea levels* | • proximity to coast/ water features<br>• location on a floodplain | • location on a floodplain | • location on a floodplain |
| | *Possible higher winds* | • aerodynamic building design (roof etc)<br>• tall building design<br>• use and selection of materials | • working at height<br>• use of cranes | • use and selection of materials<br>• working at height<br>• new roofs |
| | *More solar radiation* | • passive solar design<br>• passive cooling<br>• active cooling | • external working<br>• use and selection of materials | • use and selection of materials<br>• passive cooling<br>• active cooling |
| | *Greater specific humidity* | • comfort of internal building users<br>• damp problems | • use and selection of materials | • discomfort to building users<br>• health concerns from biological contaminants (mould) |
| | *More intense rainfall in winter* | • location of building/infrastructure<br>• use and selection of materials<br>• damp problems<br>• drainage design | • flooding on site<br>• use and selection of materials<br>• external working<br>• slope design | • location on a floodplain<br>• increased damp<br>• rain penetration |

### 5.3.4 Consequences of climate change on construction activities

In Chapter 4, some possible consequences of climate change for the construction industry were identified and discussed. By considering the climate change activities likely to affect a construction project, as outlined in the previous section, the possible consequences of the climate change events can be identified. The risk matrix forms a useful mechanism by which these potential consequences can be recorded and used to help in assessing the level of risk.

### 5.3.5 The overall risk assessment

For the risk assessment, information drawn from Tables 5.2 and 5.3 should be entered into the risk matrix. This enables identification of the overall level of risk from climate change events on construction activities.

Table 5.4 below provides a basis for the calculation of overall risk, taking into consideration the likelihood of climate change and the significance of the consequence to the construction industry. When calculating the overall risk to construction processes from climate change, the confidence level of the prediction is not taken into account. However, when prioritising actions and responses within an organisation, these confidence levels should be considered. For example, if two risk calculations are made and both result in moderate overall risk, it may be more appropriate to prioritise the climate change scenario with the higher confidence over the other.

**Table 5.4**    *Matrix for calculating overall risk calculations*

| Significance of consequence to construction industry | Likelihood of climate change event | | | | |
|---|---|---|---|---|---|
| | Low (0–20%) | Low (20–40%) | Medium (40–60%) | Medium (60–80%) | High (80–100%) |
| **High** | Moderate | Moderate | Moderate | High | High |
| **Moderate** | Low | Low/moderate | Moderate | Moderate | High |
| **Low** | Low | Low | Moderate | Moderate | Moderate |

## 5.4 OPTION APPRAISAL AND DEVELOPMENT OF ADAPTATION STRATEGY

Where the overall risk has been assessed as moderate or high, it will be necessary to identify how that risk can be best managed by adopting a suitable adaptation strategy. The aim should be to ensure that all residual risks meet with the objectives identified at the beginning of the risk decision-making process and, in particular, align with the risk culture within an organisation.

The following approaches could be used to help determine the adaptation strategy for each risk:

- avoiding the risk
- mitigating the risk
- funding the risk
- transferring the risk.

For each approach, various solutions will be available that will enable an organisation to manage the overall risk. The solution should be recorded in the risk matrix (see Table 5.4) and the residual risk assessed.

Some example solutions to *avoid* the risk include:

- redesign to avoid locations where flooding may occur
- planning to use materials that are not affected by climate change
- designing out moving joints.

Solutions for *mitigating* the risk might include:

- reprogramming work to avoid seasons and times of day when climate change is most likely to have an impact
- minimising the use of materials susceptible to climate change effects
- minimising the use of cranes in exposed locations
- programming to allow for the possibility of delays.

It may be possible to *manage* the risk by funding through options such as:

- self insurance
- escrow funds
- bonds.

Or risk might be *transferred* by:

- contractual agreements, which may include apportioning the risk and liabilities through indemnities and warranties
- use of insurance to transfer risk to a financially secure third-party carrier.

## 5.5 CASE STUDY

The following example draws on the information identified throughout this publication and specifically on the risk decision-making process outlined in this chapter. This scenario has been developed to provide a practical example of how climate change issues could be considered in the risk assessment process using the risk matrix provided in Appendix 3. It is not intended to represent any actual situation.

**Project background**

A contractor has been commissioned to design and build a high-rise apartment block, which the client has instructed should have a design life of 80 years. The location is in south-east England, close to a river. Construction is planned to start in the period 2005–2010.

**Risk decision process**

It has not been the intention to identify all climate change events and associated consequences in this process, merely to provide sufficient examples to demonstrate the process.

**Table 5.5** *Risk matrix for case study*

| Relevant climate change event | Confidence of prediction of change | Likelihood of climate change event* | Construction process | Possible consequence of climate change | Significance of consequence on the construction activity | Overall risk of impact | Possible adaptation strategy | Residual risk following mitigation |
|---|---|---|---|---|---|---|---|---|
| Wetter winters | High | 1% | Construction | Delays and disruption caused by higher river levels in winter[1] | High | Moderate | Plan to allow for possible flooding | Low |
| Increased winter rainfall intensity | High | 7% | Design | Penetration of cladding/windows[2] | High | Moderate | Not required | Low |
| Hotter summer days | High | 63% | Design | Effects on movement joints[3] | Moderate/ low | Moderate | Minimise number of joints | Low |
| Hotter summer days | High | 1% | Construction | Effects on concrete curing[4] | Moderate | Moderate | Programme to avoid heat | Moderate |
| Decrease in summer cloud, increase in solar radiation | Low | Not known | Design | Effects on materials of solar radiation[5] | Moderate | | Suitable materials | Moderate |
| Higher specific humidity | High | Not known | Design | Effects on materials of humidity[6] | Low | | Not required | Low |
| Drier summers | Medium | 10% | Construction | More productive work | Moderate | Moderate | n/a | |

**Notes**

\* For a construction activity the likelihood for 2020s is used, for design the likelihood for 2080s is used, both from the medium–high emissions scenario.

**Notes on specific risks**

1 The river is not tidal in this example, but is close to the estuary. If it were tidal, the likelihood of a climate change effect would be higher, as it would include a factor for likely sea level rise. As the change is most likely to be attributable to winter rain, which tends to be less intense but of longer duration than summer rain, the risk will be greater than higher up the river catchment. In this example it is possible to allow for flooding and to plan around that possibility. If that were not possible, consideration should be given to summer working, or building an access at a higher level or within a coffer dam. These latter options would have to be agreed with the Environment Agency, if they were likely to encroach into the floodplain. Note that the building itself is not located in a flood risk area (only the access to it) so design issues relating to river levels are not considered further.

2 The apartment block is designed with sealed units, so the consequence and overall risk are both assessed as low.

3 Although the consequence is not considered to be very high, the high likelihood of more frequent hot days makes this a moderate risk, requiring action to minimise the number of joints.

4 Concrete slabs are planned and so the increased likelihood of hotter days means this is a high risk. It may be possible to work at night, or to use smaller volume concrete pours, or to use a greater amount of pre-cast concrete to reduce the risk.

5 Over the design life of the building, there is likely to be much greater exposure to solar radiation. South-facing façades and the roof will need special consideration.

6 Although this risk has to be considered for the life of the building, it is considered to have a low impact, as the building will have air-conditioning

# 6 Consequences, climate change and the decision-making process

Previous chapters have highlighted the importance of identifying the consequences of climate change and have provided a methodology for integrating climate change issues into the decision-making process. This will enable climate change factors to become an integral part of planning a construction activity across the design, construction and operation stage of any project.

This chapter examines in more detail some of the impacts identified in Chapter 4 and illustrates the process of completing a risk matrix for each. The information used to complete the risk matrix has been drawn from previous sections, particularly Chapter 5.

## 6.1 DELAYS TO CONSTRUCTION PROGRAMMES

### 6.1.1 Scope

> **Identify the scope of the issue**
>
> - is the key area of concern at the construction phase, during the design of the structure, or both?

Completion to time and budget without compromising quality is a primary objective of any construction project. When projects are not completed in a timely manner, the client and contractor can suffer the financial consequences, typically in the form of increased costs.

Climate-related delays to the construction process are documented in Graves and Phillipson (2000), which describes potential implications of climate change for the built environment and includes such issues as:

- health and safety concerns with hazardous climate conditions
- potential need for sourcing alternative materials, eg water-intensive activities may need to be reduced as a result of water shortage
- site wastage through damage to materials from water, wind, sun exposure etc
- increase in mobility of soil contaminants
- actual delays to the construction process caused by unexpected climate conditions.

Setting an appropriate timescale for construction may therefore be critical to the success of the project. Programmes are already very sensitive to climatic conditions in the UK and the predicted changes in climate over the coming years may require new assumptions to be made about the duration of activities.

Unlike design activities, which may require consideration of climate-change risks many decades into the future, construction programmes are usually considered only a relatively short time (from months to one or two years) before construction begins. The climate change effects to be taken into account will therefore be more immediate and more likely to be definable. This immediacy may mean that, for a given project, the changing climate is not recognised as influencing the programme. Over time, however, the duration of component activities is likely to alter the overall programme.

## 6.1.2 Climate change context

**Identify the climate change issues**

- the climatic factors that may affect the construction process
- the climatic factors that may affect the design life of the building
- likelihood of a climate change event
- confidence in the predicted climate change scenarios

Construction is largely based on site works, which can be affected by climatic conditions. Although most of this publication considers the climate change effects possible up to the 2080s, the influence of shorter-term effects (up to the 2020s) cannot be ignored. It is these that will be of most relevance to those planning construction programmes. As noted in Section 2.3, they are generally considered more certain (if smaller in impact) than the longer-term effects. Several key climate factors have the potential to affect construction projects including rainfall, temperature, wind speed and depressions. Specifically:

- the annual volume of rainfall (precipitation) is not expected to change, but by the 2020s the UK can expect to experience a decrease in summer rainfall balanced by an increase in winter precipitation
- average summer and winter temperatures are both expected to increase, although the increase in summer temperatures is predicted to be higher than winter temperatures. The number of very hot days is predicted to rise sharply.

Table 6.1 provides a summary of the extent of these predicted climate change conditions for the UK. For detailed climate change scenarios see Table 2.1.

**Table 6.1**  *Climate change predictions relevant to on-site construction activities*

| Factor | Parameter | 2020s | Confidence |
|---|---|---|---|
| Precipitation | Average summer | 0–20% ↓ | Medium |
|  | Average winter | 0–15% ↑ | High |
| Temperature | Summer mean | 0.5–1.5°C ↑ | High |
|  | Winter mean | 0.5–1°C ↑ | Low |

↓ = decrease;  ↑ = increase

## 6.1.3 Issues affecting programmes

Extreme weather often leads to construction processes being suspended because of technical difficulties, damage to materials, problems with workmanship and for health and safety reasons. Box 6.1 provides examples of weather-related delays that have been experienced.

**Box 6.1**  *Examples of climate delays in construction processes*

- Sites may be made unworkable for extended periods because of waterlogging in winter and high river/tide levels (many sites were affected by the floods of autumn 2000)
- cranes may be forced to stand idle in high winds
- on very hot days concreting may be difficult
- water restrictions may be imposed through summer months that will affect water-reliant processes
- excavated materials may be too wet for placement with adequate compaction.

**Identify appropriate guidance**

- British standards and codes of practice
- guidance to assist with the use of British standards
- general industry guidance, eg from BRE, CIRIA
- does the guidance take into account climate change?

If a site is at risk of being flooded or waterlogged it will be prudent to make allowance in the programme for greater down-time in winter and/or maximising the activities that can be undertaken in summer.

Higher temperatures will affect the curing of materials. If this matter is not addressed effectively, it may be detrimental to serviceability through shrinkage, cracking and unacceptable finishes (see also Section 6.2). Failure to resolve these issues may necessitate repairs that will delay the project and may affect the future management of the facility.

While not strictly a programming issue, there will be a need to provide better storage of materials than has been traditional on many projects, if deterioration through exposure (rain, sunlight and heat) is to be avoided.

## 6.1.4

**Identify appropriate practices**

- past experience
- general industry guidance, eg BRE, CIRIA
- does the guidance take into account climate change?

### Current practices and guidelines

#### Programming

Logistics and planning are key to the development of budgets and schedules for any construction project. During the project planning stage careful consideration of the climate conditions that might be experienced can help ensure that these aspects are taken into account in the development of the overall risk assessment for the project. For certain climate conditions, mitigating actions may not be possible or cost-effective, but identification of the potential risk in project plans can benefit all parties involved. An important consideration for many construction projects will be the optimising of time available for earthmoving activities in the early spring and the autumn.

Factoring-in delays for short-term climate conditions or extreme events to the initial stages of the process may help to avoid large unplanned costs during a construction project. Many projects to get into difficulty when unrealistic programmes are set early on in the process. Factoring for short-term climate issues will help minimise contractors' overall risk exposure and hence lead to:

- stakeholders having greater confidence in the timely delivery of the project
- the contractor gaining a reputation for timely and on-budget projects
- fewer insurance claims and reduced premiums.

**Risk assessment process**

- is the risk a real concern?
- will there be a residual risk?
- are there any alternatives?

**Box 6.2**      *Example risk assessment checklist for project scheduling*

**Information needed**

- What are the average weather conditions at the location?
- which site activities are weather-dependent?
- when are weather-dependent activities planned relative to average weather conditions?
- where are rivers located? Is the site in a floodplain?
- have there been recent flooding events?
- what are the ground conditions? (cg clays could be of concern after extreme storm events)
- what are the groundwater conditions?

**Consequences**

- What would be the implications of programme delay?
- are there any clauses in the contract and provision in the budget that will protect clients from loss caused by delay?
- do the programme and budget include "float" provisions and how do they compare with the potential delays?
- does the project have insurance cover for damage caused by extreme climate events?
- is the risk or impact reasonably foreseeable?

## 6.1.5 Risk matrix

Depending on the type of project, several options exist for reducing the risk of any climate-related delays to the project. Table 6.2 identifies several options for reducing the risk of delays. Note that this is not meant to be a complete list of the options but rather provides examples that can help an organisation develop solutions that fit the particular project, risk of concern and its own risk culture.

**Table 6.2** *Risk matrix table – delays to construction projects*

| Relevant climate change event | Confidence of prediction of change | Likelihood of climate change event | Construction process | Possible consequence of climate change | Significance of consequence on the construction activity | Overall risk of impact | Possible adaptation strategy | Residual risk following mitigation |
|---|---|---|---|---|---|---|---|---|
| Drier and hotter summers (37% drier than average) | Medium | 10% | Construction | On-site delays caused by extreme heat | Moderate | Low | Prefabrication and off-site construction<br><br>Provide shading where appropriate<br><br>Manage delays through contractual agreements | Low |
| Wetter winters (66% wetter than average) | High | 1% | Construction | On-site delays caused by winter rains<br><br>Disruption resulting from floods | High | Moderate | Short-term projects: schedule for most appropriate season<br><br>Long-term: schedule most vulnerable aspects during most appropriate season<br><br>Construction that will deliver weather-tightness as early as possible | Low |

# 6.2 DECREASED DURABILITY AND PERFORMANCE OF MATERIALS: A REVIEW OF CONCRETE

## 6.2.1 Scope

---

**Identify the scope of the issue**

- is the key area of concern at the construction phase, during the design of the structure, or both?

---

Concrete is one of the most widely used materials in UK construction (Graves and Phillipson, 2000). In part, this is because of its durability as a construction material. If properly designed and built, concrete structures are long lasting and require little maintenance.

The use of appropriate materials and methods during the placing of concrete will increase durability and decrease vulnerability to weathering. It is important to take climate conditions into account to ensure that the right design mix and placing methods are employed for specific conditions.

## 6.2.2 Climate change context

The strength and durability that concrete possesses throughout its lifetime are influenced by the chemical and physical processes that take place during the setting and curing periods of the material, which are in turn affected by the climatic conditions in the weeks immediately following placement.

An increase in temperature and decrease in relative humidity, as is predicted for the summer months, would affect both the setting and curing periods of concrete. Durable concrete can still be achieved by adapting the concrete mix and by being more rigorous about how the concrete is placed and subsequently treated during the most critical part of the curing period. The greatest consequence of unsuitable concrete mixes, placement and curing would be increased cracking, allowing greater potential for reinforcement corrosion. Once exposed and vulnerable, the rate of corrosion of reinforcement will be accelerated in warmer and/or more humid conditions. Poorer quality and unsightly finishes on the concrete are also likely.

Predicted climate change effects of significance for concrete include:

- a substantial increase in both mean summer and winter temperatures over the next century, although the temperature increase is predicted to be higher in summer than in winter. A marked increase in the number of very hot days is also predicted

- a slight increase in annual average wind speed in the UK, although there is little certainty in the scenarios and only a long-range prediction for 2080s is available

- a slight decrease in annual relative humidity, with a high confidence that this will occur mostly in summer. Little difference is anticipated for winter.

Table 6.3 summarises the extent of these predicted climate change conditions for the UK that are of most relevance for placing and curing of concrete and that will have an impact on its durability. For detailed climate change scenarios, refer to Table 2.1.

**Identify the climate change issues**

- the climatic factors that may affect the construction process
- the climatic factors that may affect the design life of the building
- likelihood of a climate change event
- confidence in the predicted climate change scenarios

**Table 6.3**   *Climate change predictions relevant to curing of concrete*

| Factor | Parameter | 2020s | 2050s | 2080s |
|---|---|---|---|---|
| **Temperature** | Summer mean | 0.5–1.5°C ↑ | 1–3.5°C ↑ | 1.5–5°C ↑ |
| | Winter mean | 0.5–1°C ↑ | 1–2°C ↑ | 1.5–3.5°C ↑ |
| | Extreme summer temperatures | | | Approximately 10-fold increase in number of very hot days |
| **Wind speed** | Average summer | | | 11% ↓ to 13% ↑ |
| | Average winter | | | 3% ↓ to 13% ↑ |
| | Occurrence of lows (depressions below 1000 hPa) | | | ↑ in number of lows below 1000 hPa* from 5–8 annually (60% ↑) |
| **Relative humidity** | Summer | | | 0–18% ↓ |
| | Winter | | | 0–3% ↓ |

**Notes**

↓ = decrease;  ↑ = increase

\*   1000 hPa = 1000 mb = 1 standard atmosphere

As Table 6.3 shows, the UK is predicted to be subject to increased temperatures especially during the summer months. Research has shown that the 28-day compressive strength of concrete cured during high temperatures is below that of concrete cured in lower temperature conditions, assuming no changes in curing methods (APMCA, 1995).

Increases in temperatures and rate of evaporation (through lower humidity) will promote the early loss of water and could result in incomplete curing through rapid drying. It is also documented that a wind increase from zero to 15 km/h will increase the evaporation rate by approximately four times (APMCA, 1995).

## 6.2.3 Current practice and guidance

The UK, other parts of northern Europe, Asia and North America have all established techniques to ensure that maximum strength and durability of concrete structures is achieved within the current ranges of climate in those countries. Some practitioners in the UK will have experience of the techniques needed for concreting in much warmer climates such as in Arabia, but many will not. A warmer, windier and less humid climate may require some of those techniques to be adapted for the UK. Increases in temperatures and rate of evaporation are likely to promote the early loss of water and incomplete curing through rapid drying. References to guidance on concreting techniques are provided in Appendix 5.

To ensure a high-quality, durable end product, it is essential that the designer specifies appropriate concrete mixes and stipulates the conditions under which concrete should be placed, set and cured.

The ongoing changes to our climate through global warming are already being felt. Consideration should be given during design, not only to the type of concrete, but to the form and methodology for its curing if the material is to behave satisfactorily over its life.

The optimum temperature at which to cure concrete is 23°C, giving a strength up to 80–100 per cent greater than that for concrete that has not been properly cured (APMCA, 1995). This optimum temperature should be compared with some of the extremes experienced in 2003 of 38.5°C in Kent and 32.7°C in Scotland.

Surfaces cured at optimum temperatures have a decreased potential for drying, with resultant reduction in shrinkage cracking, they wear better and the water-tightness of the structure is increased. Apart from being unsightly, cracking in the concrete increases the potential for reinforcement corrosion through water ingress. Corrosion causes expansion of the reinforcement surface, leading to spalling of concrete, loss of bond and loss of strength.

**Placing concrete**

If extreme hot weather is forecast, planning the concreting for either the early morning or late afternoon and the use of cooled water may help to alleviate the effects of hotter conditions. Use of admixtures to improve workability, and increased concrete cover to reinforcement may also reduce the effects of hotter weather, but the designer needs to approve such variations. The introduction of better details and the need for improved workmanship may call for greater skills in the concreting teams and require contractors to pay more attention to the training of these, relatively lower-skilled, personnel.

## 6.2.4 Risk matrix

Concrete construction can be described in four stages: design, preparation, placing and curing. To optimise the conditions for these stages, a range of techniques can be employed for reducing the impact of extreme temperature, low humidity and wind throughout the process. The risk matrix in Table 6.4 provides an example of the different climate conditions, consequences and potential adaptation techniques to reduce risk associated with this activity.

**Table 6.4**      *Risk matrix table – laying and curing of concrete*

| Relevant climate change event | Confidence of prediction of change | Likelihood of climate change event | Construction process | Possible consequence of climate change | Significance of consequence on the construction activity | Overall risk of impact | Possible adaptation strategy | Residual risk following mitigation |
|---|---|---|---|---|---|---|---|---|
| A hot August (average temperature > 3.4°C above average) | Medium | 63% | Design | Concrete will begin to set faster, leading to loss of workability, and will dry out faster, leading to loss of strength and cracking | Moderate | Moderate | Use pre-cast concrete<br><br>Reduce construction joints<br><br>Use cement substitutes such as PFA<br><br>Use admixtures to prolong workable life of the wet concrete and reduction of initial temperatures during setting and curing<br><br>Use a mix design with a lower cement and chloride content | Low |
| | | | Construction | | Moderate | Moderate | Improve curing<br><br>Keep aggregates and mixing water cool<br><br>Wet surfaces against which concrete will be laid, to minimise moisture loss<br><br>Pour concrete without delay to avoid temperature build-up<br><br>Shade concrete to avoid direct sunlight<br><br>Lay concrete at cooler times of day, eg early morning or late afternoons<br><br>Use a low water-cement ratio to help reduce shrinkage during drying | Low |

## 6.3      POORER CONDITIONS IN THE INTERNAL ENVIRONMENT: CONSIDERATION OF INTERNAL MOULD GROWTH IN DOMESTIC DWELLINGS

### 6.3.1      Scope

**Identify the scope of the issue**

- is the key area of concern at the construction phase, during the design of the structure, or both?

The English House Condition survey (1991), suggested that around 3.2 million homes in England (some 15–20 per cent of the total housing stock) suffer from mould growth on internal surfaces. This included properties not built to modern standards. Mould causes unsightly damage to wall coatings, soft furnishings, personal effects and clothing. More importantly, it can also affect the occupants' health. People prone to asthmatic or allergic conditions (about 10 per cent of the population) may be affected by long-term exposure to concentrations of airborne spores.

Mould is a fungal growth and is directly related to the presence of dampness within a building. It typically occurs when cellulose-based materials such as plasterboard, wood framing and insulation become damp. Causes include rising damp, rain penetration and condensation. The difference between each issue is highlighted in Box 6.3. The text following the box focuses on mould growth associated with rain penetration.

**Box 6.3**      *Causes of damp problems associated with mould growth within internal environments*

**Rain penetration** – normally caused by rainwater coming directly through the walls or roofs, or at junctions with openings and poor detailing at lintels. It appears as a damp patch on the inside wall. Usually occurs in localised patches with well-defined edges, but not in any particular position. Patches will increase in wet weather, especially after driving rain.

**Rising damp** – caused by water rising upward through a permeable wall structure because of the absence or failure of a damp-proof course. The water rises through capillary action in the pores of the masonry. Horizontal tidemarks usually result, with a well-defined edge about 600–900 mm above ground level on external walls. Appearance will remain unchanged for long periods.

**Condensation** – occurs when moist air comes into contact either with cooler air or, more usually, with a surface at a cooler temperature. It normally results from living styles, environment and lack of ventilation, rather than building problems, although the types of materials used in buildings are often a contributory factor.

### 6.3.2      Climate change context

**Identify the climate change issues**

- the climatic factors that may affect the construction process
- the climatic factors that may affect the design life of the building
- likelihood of a climate change event
- confidence in the predicted climate change scenarios

Mould essentially requires food, oxygen, water and a suitable temperature to be able to grow. The most important factor associated with the presence of mould is water, in the form of moisture, but other climatic inter-relationships required to produce ideal conditions for mould growth are:

- ventilation
- thermal insulation
- internal heating
- surface absorption.

The predicted warmer and drier climate in the UK, particularly during the summer months may decrease the optimum conditions associated with climate-related mould growth. However, increased rain penetration consequent on higher rainfall intensity, including driving rain predicted during the winter months, may negate any of the positive effects of a warmer climate.

Predicted climate change effects of significance for damp and mould problems include the following:

- the annual volume of rainfall is not expected to change, but by the 2020s the UK can expect summer rainfall to decrease and winter rainfall to increase

- a predicted slight increase in the annual UK average wind speed. There is little certainty in the scenarios, however, and only a long-range prediction for 2080s is available. Summer winds are expected to be more variable, ranging from an 11 per cent decrease to a 13 per cent increase, while winter variations will be less. The number of depressions is predicted to increase only slightly over winter, but this brings the potential for more driving wind and rain

- an expected substantial increase in both mean summer and winter temperatures during the century, although the increase in summer temperatures is predicted to be higher than the winter.

Table 6.5 provides a summary of the extent of those predicted climate change conditions for the UK that are most relevant to conditions for internal mould. For detailed climate change scenarios see Table 2.1.

**Table 6.5** *Climate change predictions relevant to mould growth in the internal environment*

| Factor | Parameter | 2020s | 2050s | 2080s |
|---|---|---|---|---|
| Precipitation | Average summer | 0–20% ↓ | 0–40% ↓ | 0–60% ↓ |
| | Average winter | 0–15% ↑ | 0–25% ↑ | 0–40% ↑ |
| | Extreme winter precipitation | | | 100% ↑* |
| Temperature | Summer mean | 0.5–1.5°C ↑ | 1–3.5°C ↑ | 1.5–5°C ↑ |
| | Winter mean | 0.5–1°C ↑ | 1–2°C ↑ | 1.5–3.5°C ↑ |
| Wind speed | Average summer | | | 11% ↓ to 13% ↑ |
| | Average winter | | | 3% ↓ to 13% ↑ |
| | Occurrence of lows (depressions below 1000 hPa) | | | ↑ in number of lows below 1000 hPa from 5 to 8 annually |

**Notes**

↓ = decrease;  ↑ = increase

\*   Medium-high emissions only

## 6.3.3

**Identify the climate change impacts on the project**

- what will be the consequence of the climate change impact?
- do I need to take account of climate change in the design of the structure?
- do I need to take account of climate change in the construction phase?
- are there any alternative solutions?

### Current practice and guidance

The generation of mould from water penetration is primarily a problem affecting older building stock and should not occur in newer buildings. Older buildings are more likely to have solid walls that have become permeable, allowing rain penetration, or deteriorated seals with the same result. Rainfall penetration is much less common in the cavity walls commonly found in modern buildings. Missing roof tiles, damaged rainwater goods, or damaged render or cladding may be other contributors, as could the particular exposure of the building.

To minimise the potential for rain penetration into a building and development of internal mould, the construction materials and methods utilised need to be robust and appropriate for the exposure zone. BS 8104:1992 identifies a calculation methodology for assessing this exposure risk across the UK. Further to this, the BRE Good Repair Guide 33 (BRE, 2002) highlights the increased risk from potential climate change events, and suggests that the zone should be upgraded when calculating the exposure limit: ie if calculated as zone 3, one should upgrade to zone 4 exposure.

In the USA builders have been sued on the basis of the presence of mould in new houses and for failing to ensure that the building was adequately protected against rain penetration (Marsh, 2003). If rain penetration continues to be a problem in new buildings in the UK, not only would current health-related risks continue to be evident, but there may also be an associated increase in insurance claims and liabilities.

Preventing the development of mould is reliant on minimising or removing the conditions under which the mould will grow. The Building Regulations 2000 and the associated approved documents include provisions for the preventing moisture problems in buildings (see Appendix 5).

An increase in intense rainfall could make rain penetration the biggest driver of increased mould development resulting from climate change. The building fabric and such attachments as external louvres should be designed to prevent the ingress of driving rain. Material that can trap both dirt and moisture, such as fibrous media found in filters, sound attenuators and duct linings, should not be situated close to louvres. Other external considerations include the detailing of the external fabric, which should take into account the potential for wind-driven rain.

Remedies are available for protecting existing buildings from predicted increases in rain penetration. They include repointing, applying water repellent to masonry, and painting of rendering and cladding (see Appendix 5 for references to guidance).

Rectifying the problem of internal mould is usually reliant on identifying and sealing the source of the moisture, not just the removal of the mould itself. If moisture remains, mould can recolonise within 24–48 hours.

## 6.3.4

**Risk assessment process**

- is the risk a real concern?
- will there be a residual risk?
- are there any alternatives?

### Risk matrix

The risk matrix in Table 6.6 provides an example of the different climate conditions, consequences and potential adaptation techniques to reduce risk associated with internal mould.

**Table 6.6**    *Risk matrix table – internal mould resulting from rain penetration in high wind-driven rain exposure zone*

| Relevant climate change event | Confidence of prediction of change | Likelihood of climate change event* | Construction process | Possible consequence of climate change | Significance of consequence on the construction activity | Overall risk of impact | Possible adaptation strategy | Residual risk following mitigation |
|---|---|---|---|---|---|---|---|---|
| Wetter winters (66% wetter than average) | High | 7% | Maintenance | Rain penetration through existing solid wall due to absent or degraded vapour barrier, or increased porosity | Moderate | Low | Repointing mortar | Low |
| | | | Design | Penetration through window and door joints due to poor detailing | Moderate | Low | Conformance with Building Regs Approved Document C | Low |
| | | | Design | Penetration through ventilation openings | Moderate | Low | Instal multi-stage louvres | Low |

## 6.4    SUBSIDENCE AND HEAVE PROBLEMS IN DOMESTIC DWELLINGS

### 6.4.1    Scope

Subsidence and heave can occur as a result of movement within a variety of soil types, although the primary cause of these movements in the UK is the shrinking and swelling of clay soils. Movements caused by the shrink/swell cycle can seriously damage buildings, producing cracks in the walls (internal and external) and sometimes resulting in structural damage. The soils posing the greatest risk mostly occur in the south and east of England, where highly plastic clay predominates. The major problems have been within the domestic housing industry, as commercial buildings typically have much deeper foundations.

### 6.4.2    Climate change context

Changing climatic conditions such as higher temperature, less rainfall and consequent reduced groundwater will cause highly plastic clay soils to shrink, while heavy rainfall will have the opposite effect, as the increased soil moisture causes clays to swell. The key climatic factors relating to subsidence and heave are winter rainfall, increased temperature and drier summer conditions, characterised as follows:

- overall the total annual volume of rainfall is not expected to change significantly, but the UK can expect that by the 2020s there will be an increase in winter rainfall and a corresponding decrease in summer
- both mean summer and winter temperatures are expected to rise, but the increase in summer temperatures is predicted to be higher than that in winter. The number of very hot days is expected to rise sharply.

Table 6.7 summarises the extent of the predicted changes in climate conditions for the UK that are most relevant to subsidence. For detailed climate change scenario information see Table 2.1.

**Table 6.7**     *Climate change predictions relevant to subsidence and heave*

| Factor | Parameter | 2020s | 2050s | 2080s |
|---|---|---|---|---|
| Precipitation | Average summer | 0–20% ↓ | 0–40% ↓ | 0–60% ↓ |
| | Average winter | 0–15% ↑ | 0–25% ↑ | 0–40% ↑ |
| | Extreme winter precipitation | | | 100% ↑ |
| Temperature | Summer mean | 0.5–1.5°C ↑ | 1–3.5°C ↑ | 1.5–5°C ↑ |
| | Winter mean | 0.5–1°C ↑ | 1–2°C ↑ | 1.5–3.5°C ↑ |
| | Extreme summer temperatures | | | Approx 10-fold ↑ in number of hot days |

**Notes**

↓ = decrease;  ↑ = increase

\*   Medium-high emissions only

It is predicted that the south and east of England will be subject to the driest summers. As a consequence, and taking account of the predominance of clay soils in this geographic region, the problem of subsidence in this area could become even more pronounced. Recent research has identified an increase in subsidence-related insurance claims during drought conditions previously experienced in the UK (Graves and Phillipson, 2000).

## 6.4.3     Current practice and guidance

> **Identify appropriate guidance**
>
> - British standards and codes of practice
> - guidance to assist with the use of British standards
> - general industry guidance, eg from BRE, CIRIA
> - does the guidance take into account climate change?

The susceptibility of a building to cracking as a result of ground movement depends on various factors, the most important of which are the depth of the foundations and the method of construction.

**New build**

Before new domestic structures are built, a ground investigation is required so that suitable foundations can be designed. Current practice requires that these investigations should include consideration of:

- geographical location
- potential volume change in the clay soil (from both wetting and drying)
- tree species, location and height (trees absorb moisture from the ground, exacerbating the normal swell/shrink cycle; if they are cut down for development, the moisture will return, initiating greater swelling in clay soils).

The NHBC has identified regional climatic zones for the UK (NHBC, 2003a), which illustrate a general reduction in the minimum foundation depth required from south-east to north-west. However, no future climate-related scenarios have been incorporated into these recommendations. Little guidance is available on the influence that climate change may have on foundation design.

The report *Potential implications of climate change in the built environment* (Graves and Phillipson, 2000) highlights some of the approaches that may be needed to allow for climate change effects:

- deeper foundations for new developments to avoid damage caused by foundation movements

- increased depth of foundations near to trees, possibly including piles and ground beams for the worst cases

- possible development of new foundation technologies, such as prefabricated strips or pile and beam foundations.

In addition, this study estimates that there would be an annual cost increase of around £32 million to comply with the assumption that by 2080 minimum foundation depths on clay soils will have to increase by 0.5 m.

The effect of drier summers and warmer temperatures combined with wetter winters is likely to have an appreciable effect on shrinkable soils, with greater extremes in the shrink/swell cycle. It is also possible that these combined effects will start to affect soils that have not previously been considered as susceptible to the shrink/swell cycle. As noted, the presence of trees will have a considerable influence on the occurrence of subsidence, and the water supply during the summer months may become even more restricted. Designers will therefore need to consider the potential changes in rainfall over the design life of the structure and the suitability of type and depth of foundations required.

**Existing build**

Since the inception of the NHBC foundation guidelines, houses built in the past 30 years in areas of shrinkable soils with a modified plasticity index of at least 10 per cent are likely to have been constructed with minimum foundation depths of 0.75 m. The 10-year NHBC insurance for accredited houses indicates that NHBC considers the guidelines sufficient to accommodate the potential risk of subsidence and heave. The minimum foundation depth increases when the effect of trees, both existing, proposed and removed, is taken into account in accordance with the requirements of Chapter 4.2 of *NHBC standards*.

Domestic dwellings currently considered to be at high risk of subsidence and heave include houses built between the 1920s and 1970s, when shallow foundations of less than 0.45 m tended to be used (Driscoll and Crilly, 2000).

Insurers have noted a strong linkage between longer dry periods and an increase in insurance claims related to subsidence and heave. This tends to confirm that the drier summer conditions predicted in the climate change scenarios could shift the problem of subsidence and heave beyond those high-risk dwellings identified above. The FBE report (Driscoll and Crilly, 2000) estimates that insurance claims from subsidence and heave will have increased from £400 million in 2000 to between £600 million and £800 million by 2080 (based on current values).

To identify a remedy, it is essential to be able to identify the cause of subsidence and heave in existing structures, such as the presence of trees and the type of foundation supporting the structure. This will ensure that the consequence and the solution reflect the actual risk. Table 6.8 identifies several remedial methods for subsidence and heave in existing structures and the appropriateness of use. For guidance on the identification and severity of subsidence and heave problems, refer to Appendix 5.

**Table 6.8** *Remedial methods for subsidence and heave*

| Remedy | Appropriateness |
|---|---|
| Do nothing | • When damage is low, eg cosmetic cracks<br>• If cause is well established, such as tree issue in times of extreme weather<br>• If situation will not worsen |
| Tree removal/lopping | • When advised by qualified arboriculturalist<br>• When foundations have been deepened by underpinning |
| Underpinning | • To stabilise the building |
| Structural repair | • Repairing damage at point of weakness |

## 6.4.4 Risk matrix

The risk matrix, Table 6.9, provides an example of the different climate conditions, consequences and potential adaptation techniques to reduce the risk associated with subsidence.

**Table 6.9** *Risk matrix table – subsidence in domestic dwellings*

| Relevant climate change event | Confidence of prediction of change | Likelihood of climate change event | Construction process | Possible consequence of climate change | Significance of consequence on the construction activity | Overall risk of impact | Possible adaptation strategy | Residual risk following mitigation |
|---|---|---|---|---|---|---|---|---|
| Drier summers (37% drier than average) | Medium | 50% | Design | Subsidence of houses on clay soils | Moderate | Moderate | Adequate ground investigation<br>Ensure adequate foundation | Low |
| Wetter winters (66% wetter than average) | High | 7% | Design | Subsidence of houses caused by tree growth | Moderate | Low | Ensure adequate drainage<br>Include management plan for control of tree growth | Low |

## 6.5 SLOPE INSTABILITIES

### 6.5.1 Scope

Slope instability is a common global phenomenon that has resulted in failures ranging in scale from minor slips to major mass down-slope movements causing severe damage to property and loss of life.

Numerous landslides have occurred in the UK, although the majority have caused only minor damage or inconvenience. Nevertheless, there have been catastrophic incidents resulting in loss of life, such as the Aberfan disaster, when 144 people, 116 of them children, were killed after a tip of coal waste collapsed in South Wales in 1966.

Slopes fail when rock, soil or waste becomes unstable and moves downwards under the influence of gravity. A common trigger of this instability is the build-up of water pressures from intense rainfall events or prolonged periods of rain. It has been known to cause damage and destruction to property, economic loss through disruption of transport networks, and delays to the construction process. In some instances, as noted above, it has caused loss of life. Box 6.4 provides an overview of issues associated with slope instability in the UK.

**Box 6.4**      *Slope instability in the UK – an overview*

Slope instability and resulting landslides can cost millions of pounds annually through the temporary closure of roads and delays to construction projects. It was estimated that repairs to roads in North Wales from previous landslips cost £2 million in 2003.

Within the UK there are extensive areas of naturally occurring landslides, particularly in the Scottish Highlands, Cotswolds, Pennines, South Wales and coastal districts in Dorset and the Isle of Wight. During dry weather these slopes are generally stable, but their latent instability can be reactivated by construction work and heavy rainfall.

It is not just natural slopes that may be at risk. Transportation infrastructure has associated embankments and cuttings; there are flood defence embankments, and soil bunds are increasingly being used for environmental screening and noise abatement measures. All of these slopes may be at risk if not designed against future environmental changes.

## 6.5.2      Climate change context

*Identify the climate change issues*

- the climatic factors that may affect the construction process
- the climatic factors that may affect the design life of the building
- likelihood of a climate change event
- confidence in the predicted climate change scenarios

Both natural and human influenced events can contribute to the instability of slopes. Water is recognised as being a primary contributor to the instability of a slope through the increase of pore water pressure within soils and rocks, and it also contributes to the mobility of debris down the slope.

Slopes are only seriously at risk of failure if there is an incipient slip zone with a safety factor little more than 1.0. In these circumstances a tension crack or cracks will tend to open at the top of the potential failure zone. Increased temperature and drier conditions will exacerbate shrinkage and cause the crack to widen, particularly within cohesive soils. Sudden intense storms, or periods of increased rainfall in winter, allow the tension cracks to fill with water, and this alone may be enough to trigger failure. Winter rainfall also raises groundwater levels and pore pressures within the slope, which adds to the stresses in the potential failure zone. The whole wetting and drying cycle is a continual process that affects the stability of slopes and may become more crucial as the cycle becomes more extreme under a changing climate.

Slopes engineered to modern standards with adequate safety factors are unlikely to become unstable as a result of changes to the climate.

Natural landslips that may affect a development need careful investigation and may require remedial action against failure. Climate change considerations affecting slope stability include the following:

- the total annual volume of rainfall is not expected to change significantly, but the UK can expect by 2020s that winter rainfall will increase with a corresponding decrease in summer rainfall
- both mean summer and winter temperatures are expected to increase, although the increase in summer temperature is predicted to be higher than that for winter. The number of very hot days is expected to rise sharply.

Table 6.10 summarises the extent of the predicted changes in climate conditions for the UK that are most relevant for slope instabilities. For detailed climate change scenario information see Table 2.1.

**Table 6.10**   *Climate change predictions relevant to slope instability*

| Factor | Parameter | 2020s | 2050s | 2080s |
|---|---|---|---|---|
| Precipitation | Average summer | 0–20% ↓ | 0–40% ↓ | 0–60% ↓ |
| | Average winter | 0–15% ↑ | 0–25% ↑ | 0–40% ↑ |
| | Extreme winter precipitation | | | 100% ↑* |
| Temperature | Summer mean | 0.5–2°C ↑ | 1–4°C ↑ | 1.5–5°C ↑ |
| | Winter mean | 0.5–1°C ↑ | 1–2.5°C ↑ | 1.5–4°C ↑ |
| | Extreme summer temperatures | | | Approx 10-fold ↑ in number of hot days |

**Notes**

↓ = decrease;   ↑ = increase

\*   Medium-high emissions only

**Identify appropriate guidance**

- British standards and codes of practice
- guidance to assist with the use of British standards
- general industry guidance, eg from BRE, CIRIA
- does the guidance take into account climate change?

The predicted increase in rainfall intensity particularly during the winter months may have a marked influence on the incidence of slope instability throughout the UK. Short periods of intense rainfall have also been known to result in instabilities primarily associated with shallow slips and debris flows. This type of instability tends to be localised and dependent on the geology, climate and geomorphology of each region.

Significant intense rainfall events have been recorded in the UK. For example, in August 2003 Carlton in Cleveland, Yorkshire, experienced 48 mm of rainfall in 15 minutes (Met Office, 2003).

Box 6.5 identifies landslips in the UK that have occurred after intensive or prolonged periods of rain.

**Identify the climate change impacts on the project**

- what will be the consequence of the climate change impact?
- do I need to take account of climate change in the design of the structure?
- do I need to take account of climate change in the construction phase?
- are there any alternative solutions?

**Box 6.5**   *Examples of landslips after intensive or prolonged periods of rain*

- During 1993 several houses in Bolsover were damaged and subsequently demolished after a prolonged period of rain reactivated an ancient landslide
- also in 1993, the Holbeck Hall Hotel, on South Cliff, Scarborough, Yorkshire, collapsed after a cliff receded more than 60 m, again after a prolonged period of rain*
- in 2000, after an extended period of rain, cliff falls and landslides were widespread in Ventor Undercliff, Isle of Wight, while other houses were destroyed and gardens and foot-paths blocked or destroyed.*

\*   Both these events were natural cliff failures, exacerbated by rainfall.

**Source:** PPG 14 *Development on unstable land*, Annex 1.

## 6.5.3 Current practice and guidance

Failures within engineered earthworks have occurred over many years. The introduction of asset management systems by infrastructure owners, such as the Highways Agency and Network Rail, has led to the identification of more failures, leading to increased investigation, monitoring and remediation. Non-transport-related construction works could use the additional knowledge and experience from these new systems to good effect when assessing the stability of slopes.

It is probable that climate change factors will bring about greater numbers of failures, with the most likely cause being the greater extremes in the wetting and drying cycle.

Where a previous landslide has occurred, or where construction is planned, geotechnical engineers should be employed to investigate the ground conditions, to assess the risk of slope instability and to undertake a design that prevents loss of property and life. The potential for instability will be affected by geology, groundwater regime, slope profile and other external factors.

Methods for rectifying instabilities in existing slopes include reducing slope angles, removing failed materials and replacing with engineered fill, introducing surface and counterfort drains, and constructing retention or stability measures. Given the large number of slopes in the UK's existing transport infrastructure it would be reasonable to expect an increase in slope failures as climate changes progress.

**Remediating and mitigating slope instability**

There are many methods that can be adopted in the repair and remediation of earthwork slopes. These methods can be split into the following categories, based upon CIRIA publications C591 (Perry and Brady, 2003) and C592 (Perry *et al*, 2003):

- rebuild techniques
- retaining structures
- drainage measures
- *in-situ* reinforcement
- grouting
- surface protection.

Retaining structures constructed at the toe of a slope can allow the slope angle to be reduced. The reconstruction of failed slopes with geotextile reinforcement has become a popular and less expensive solution, especially given legislative restrictions on disposal of materials and associated costs. The provision of additional drainage, especially within the slope, removes water from slopes that may be considered at risk of failure.

The planting of vegetation with deep, stabilising roots is known to be an effective mitigation technique (CIRIA PR81 – Greenwood *et al*, 2001). The vegetation has the effect of removing water from the soils and increasing the *in situ* strength stability.

Nevertheless, even with the presence of deep-rooted trees, slopes have become unstable during periods of intense rainfall.

**Suggested method for assessing risks from natural slopes adjacent to new developments**

Before starting development on or beside natural slopes, a geological and geomorphological assessment of the landslip potential should be undertaken. This is an inherent part of current design philosophy and should be undertaken by an experienced geotechnical engineer or engineering geologist. The assessment should include:

- consultation of the national landslides database; analysis of geological maps

- enquiries to other authorities, including local authorities, the Highways Agency etc, which often have additional information on high-risk landslide areas

- an initial engineering assessment of the potential for instability.

**Box 6.6**    *Checklist for determining issues for slope instability*

---

**Key questions**

- has there been a history of slope failures in the locality of the site?
- is there evidence of previous failures and existence of former slope stability measures?
- what are the general geology and soil/rock conditions on site and are they prone to instability?
- what are the natural slope features and what features are being introduced?
- is the vegetation cover protecting the slope or contributing to instability?
- what is the slope geometry, scale and gradient?
- what is the site's groundwater and hydrology?

**Consequence of event**

- what features are down-gradient of slopes?
- what features are on top of the slopes?
- are the slopes in "use"?

---

**Risk assessment process**

- is the risk a real concern?
- will there be a residual risk?
- are there any alternatives?

Box 6.6 outlines typical issues that should be considered before development starts. In any circumstance where slope instability is considered possible, a full engineering assessment, which may need to include a site investigation, should be undertaken and necessary designs prepared.

See Appendix 5 for details of current guidance for development in areas of unstable land.

### 6.5.4    Risk matrix

The risk matrix, Table 6.11, provides an example of the different climate conditions, consequences and potential adaptation techniques to reduce risk associated with slope instability.

**Table 6.11**    *Risk matrix table – slope instability*

| Relevant climate change event | Confidence of prediction of change | Likelihood of climate change event | Construction process | Possible consequence of climate change | Significance of consequence on the construction activity | Overall risk of impact | Possible adaptation strategy | Residual risk following mitigation |
|---|---|---|---|---|---|---|---|---|
| Wetter winters (66% wetter than average)<br>More intense storms in summer | High | 7% | Design | Failure of previously stable clay slopes | Moderate | Moderate | Adequate ground investigation<br>Avoid over-steep slope<br>Ensure adequate drainage<br>Stabilise slope | Low |

## 6.6  DAMAGE TO THE FABRIC OF BUILDINGS: POTENTIAL EFFECTS ON CLADDING

### 6.6.1  Scope

Cladding provides the outer skin of a framed structure and so is largely responsible for resisting the penetration of weather into the internal environment. Modern structures make use of a wide variety of cladding methods, from brick and block panels to modern glass curtain walling systems. These are the most exposed elements of the structure and bear the full effect of the external climatic conditions, including temperature, wind, rain and sunlight.

### 6.6.2  Climate change context

Systems and materials forming the external skins of buildings have been designed to resist the climate conditions prevalent in the UK. These systems have been developed empirically to satisfy local requirements and to achieve the required design life. If the climate changes, these empirically derived systems may either fail prematurely or have a shorter service life. Critical to the issue is the degree of change that can be expected.

Predicted climate change effects that have a bearing on cladding include the following:

● annual volume of rainfall is not expected to change, but by 2020s the UK can expect summer rainfall to decrease while winter rainfall will increase

● annual average wind speed is predicted to increase slightly in the UK, but there is little certainty in the scenarios and only a long-range prediction for 2080 is available. The number of depressions is predicted to increase only slightly during the winter, although more driving wind and rain may be experienced

● both mean summer and winter temperatures are expected to rise substantially over the next 80 years, although the increase in summer temperature is predicted to be higher than that in winter. The number of very hot days is expected to rise markedly.

Table 6.12 summarises the extent of the predicted changes in UK climate conditions that are most relevant for cladding. For detailed climate change scenarios see Table 2.1.

**Table 6.12**  *Climate change predictions relevant to potential effects on cladding*

| Factor | Parameter | 2020s | 2050s | 2080s |
|---|---|---|---|---|
| Precipitation | Average summer | 0–20% ↓ | 0–40% ↓ | 0–60% ↓ |
| | Average winter | 0–15% ↑ | 0–25% ↑ | 0–40% ↑ |
| | Extreme winter precipitation | | | 100% ↑* |
| Wind speed | Average summer | | | 11% ↓ to 13% ↑ |
| | Average winter | | | 3% ↓ to 13% ↑ |
| | Occurrence of lows (depressions below 1000 hPa) | | | ↑ in number of lows with 1000 hPa from 5 to 8 annually (60% ↑) |
| Temperature | Summer mean | 0.5–1.5°C ↑ | 1–3.5°C ↑ | 1.5–5°C ↑ |
| | Winter mean | 0.5–1°C ↑ | 1–2°C ↑ | 1.5–3.5°C ↑ |
| | Extreme summer temperatures | | | Approx 10-fold ↑ in number of very hot days* |

\*   Medium-high emissions scenario available only;  ↓ = decrease;  ↑ = increase

Commercially there will always be pressure to provide the most cost effective solution. This means that cladding, like all other components, has to be value engineered to suit a specification and a particular location. If the ranges of precipitation and temperature for which the cladding has been designed change, then this may render the chosen system unsuitable.

## 6.6.3    Current practice and guidance

Codes of practice and British Standards indicate that the effective life of a structure is around 50–60 years, depending on the type of structure. Curtain walling and non-masonry cladding will typically have a life of about 20–25 years, reducing to around 10 years for gaskets and sealants. With this in mind, the cladding standards need to be upgraded to suit the predicted change in conditions affecting the service life of the element. Relevant guidance is included in Appendix 5.

As with all potential changes, the earlier in procurement that this issue is addressed the better, as the cost of enhancing the specification is more likely to be accommodated. Of potentially far greater financial significance is loss of service life leading to unplanned maintenance and a corresponding reduction in the time to replacement, especially as there are likely to be costs associated with the loss of use or limitations in use of the building during repairs. Further information on refurbishment and enhancement of building specifications is available in CIRIA Report 133 (Perry, 1994) and in CIRIA C621 (Fawcett and Palmer, 2004).

### Design issues

*Identify the climate change impacts on the project*

- what will be the consequence of the climate change impact?
- do I need to take account of climate change in the design of the structure?
- do I need to take account of climate change in the construction phase?
- are there any alternative solutions?

The frequency of joints in cladding systems has been developed from historical and measured events and so has been based on existing conditions. The predicted rise in temperatures will lead to an increase in the thermal range to which cladding will be exposed, resulting in a need for more or larger joints to control the thermal expansion. This in turn is likely to demand improved performance from the joints and the sealants. These can be crucial elements in the weather-tightness of the system. New or modified cladding systems may require separate rain screening tests to confirm their water-tightness.

Sealants are also subject to deterioration from climatic conditions, particularly in increased temperatures and UV light exposure. Additional weathering data may need to be sought from materials suppliers to confirm suitability for the assumed design climate conditions. Alternative designs, such as revised joint configurations to shield the sealant from UV light exposure, might need to be employed to address this risk.

Many sheet cladding systems are colour-coated. The current formulation of coatings is designed to achieve the required design life specification, but weathering data against increased UV exposure may be needed to confirm continuing suitability.

Common cladding panels on modern buildings are either framed as part of a cladding system or are restrained back to the main structural frame. The restraint to these panels may also need to be addressed for structural adequacy. Although predictions of increased wind speeds are uncertain and available only for the 2080s, the risk of the resulting increased forces may, for critical structures, require these restraints to be strengthened. Alternatively, it may be acceptable to recognise a reduction in service life with increased maintenance costs. Because of the uncertainty of future wind speeds, readers are encouraged to review UKCIP information from time to time for further predictions.

**Maintenance issues**

Multi-storey buildings are subject to the greatest exposure, and access is rarely easily available for major maintenance works. Cleaning cradles used for routine maintenance are not appropriate for wholesale repair or re-cladding. If climate change were to cause premature failure of the cladding system, large sums would have to be added to maintenance budgets to provide replacement cladding.

**Conclusions**

All fixing systems for cladding are designed with factors of safety that make allowance for variations in materials and workmanship. It is unlikely that a properly specified and fixed system would be rendered unsafe by an increase of 10–15 per cent in wind velocity and hence pressure, but issues may become apparent where the cladding has been either underspecified or where the fixing is less than perfect.

By the same token, design warranties provided for proprietary cladding systems will be based on current standards and statutory requirements. Designers should not rely on a reduction in safety factor to overcome potential increases in wind velocities, as this would be likely to invalidate such warranties.

## 6.6.4    Risk matrix

---

**Risk assessment process**

- is the risk a real concern?
- will there be a residual risk?
- are there any alternatives?

---

The risk matrix, Table 6.13, provides an example of the different climate conditions, consequences and potential adaptation techniques to reduce risk associated with cladding failure.

**Table 6.13**    *Example risk assessment matrix – effects of climate change on cladding*

| Relevant climate change event | Confidence of prediction of change | Likelihood of climate change event | Construction process | Possible consequence of climate change | Significance of consequence on the construction activity | Overall risk of impact | Possible adaptation strategy | Residual risk following mitigation |
|---|---|---|---|---|---|---|---|---|
| Wetter winters (66% wetter than average) | High | 7% | Design | Damage to façade and potential water penetration through cladding not coping with increased intensity in rainfall | Moderate | Low | Enhance the exposed finish of the cladding to allow for increased rainfall intensity<br><br>Potential to redesign cladding joints to allow for drained sections with the sealant located inboard | Low |
| Increase in number of depressions per season in winter (baseline of 5 per season) | Low | 63% | Design | Damage to joints as sealants and gaskets not rigid enough to cope with increased wind driven rain from depressions | Low | Moderate | Sealants and gaskets enhanced with the structural framework stiffened to reduce flexure in storms | Low |
| | | | Design | Glazing not strong enough to cope with increased wind pressures | Low | Moderate | Increase thickness of glazing to reduce wind forces or reduce panel sizes | Low |

## 6.7      STRUCTURAL DAMAGE FROM WIND-RELATED EVENTS

### 6.7.1      Scope

Generally, the taller the structure the more significant the effect of any increase in wind forces, although significant risks are not limited to tall structures. Increases in wind velocity, duration and frequency can affect not just structural design assumptions and stresses, but also produce occupancy issues such as sway effects.

Confidence within the climate change models of the likely changes in wind speeds is generally low, but the consequences of wind damage can be significant, with high resultant costs.

### 6.7.2      Climate change context

Climate changes that have been predicted for the UK include the possibility of increased wind speeds, along with greater gust duration and with storms appearing with greater frequency. Current structural design codes and standards are based on climatic conditions that have traditionally been experienced in the UK. If the predicted increase in average wind speed occurs, then structures designed to current standards may be subject to overstressing from wind loading during their lives.

The key climatic factors that relate to structural damage from wind-related events are average wind speed and an increase in storms (depressions or lows).

Annual average wind speed in the UK is predicted to increase slightly, but there is little certainty in the scenarios and only a long-range prediction for the 2080s is available. Within this small overall increase, the prediction for summer wind speeds varies from an 11 per cent decrease to a 13 per cent increase, while that for winter is less variable. The number of winter depressions is predicted to increase from five to eight, which may increase the incidence of driving wind and rain.

Table 6.14 provides a summary of the predicted most relevant to wind loading. For detailed climate change scenario information see Table 2.1.

**Table 6.14**     *Climate change predictions relevant to structural damage*

| Factor | Parameter | 2020s | 2050s | 2080s |
|---|---|---|---|---|
| | Average summer | | | 11% ↓ to 13% ↑ |
| | Average winter | | | 3% ↓ to 13% ↑ |
| **Wind speed** | Occurrence of storms (lows or depressions below 1000 hPa) | | | ↑ in number of winter lows below 1000 hPa from 5 to 8 annually (60% ↑)* |

**Notes**

↓ = decrease;  ↑ = increase

\*  Medium-high emissions only

## 6.7.3    Current practice and guidance

### Existing structures

The construction of older, traditional structures is generally heavier than that of their modern equivalents. These older structures in masonry or concrete are less likely to be at risk from the effects of increased wind forces as they are inherently more rigid. Although individual elements may be vulnerable, the integrity of the structures overall is unlikely to be an issue.

The most difficult and costly problems can be expected to arise from having to strengthen major modern structures retrospectively to resist the effects of increased wind forces. Rarely will strengthening be impossible from a technical standpoint; it will generally be an issue of cost-effectiveness.

Some of the issues that need to be considered in strengthening existing buildings against increased wind loading include:

- how to strengthen foundations, particularly beneath key structural elements such as shear cores

- access arrangements for construction plant for additional piles

- the potential magnitude of sway effects, and the control of accelerations or resonant frequencies (likely to be a particular issue for long-span bridges)

- loss of usable space to strengthening, bracing or damping requirements

- the need for temporary works during strengthening

- loss of use, or loss of revenue, during strengthening works

- strengthening of cladding, particularly modern glass curtain walling. The potential effects of climate change on cladding is discussed in Section 6.6.

### New build

Both CP3:Chapter V-2:1972 and its replacement BS 6399-2:1997 *Loading for buildings. Code of practice for wind loads* include wind-velocity contour maps of the UK (Isopleth contours). They form an intrinsic part of the design process, as the forces that structures of all kinds are designed to resist are based on these wind speed contours. While provision is made for uncertainties and variations in the forces arising, no specific allowance has been incorporated into the British standard for progressive changes in wind speed and there is currently little discussion about the influence that climate change factors may have on wind loading. Given the lack of confidence in climatic predictions of wind speed, this may not be surprising, but the subject may need to be revisited as predictions become more definitive.

The British standards and supplementary BRE digests have identified the importance of such subtleties in obtaining more realistic values of wind loading by incorporating further parameters into the calculations. These parameters include the aspect of the building, design of external façade, proximity to slopes and hillsides, and location and situation relative to the coast. Standards can therefore be supplemented with parameters for climate change effects once more accurate information is available. The current guidance is identified in Appendix 5.

During the construction phase, projects are heavily dependent on the use of cranes. As mean wind speed increases, the percentage of occasions when it may be too windy for safe use of cranes will increase.

### 6.7.4 Risk matrix

**Risk assessment process**

- is the risk a real concern?
- will there be a residual risk?
- are there any alternatives?

Depending on the type of project, there are options for reducing the risk of any climate-related issues associated with wind loading (Table 6.15). The table is not intended as a complete list of the options, but to provide examples and thus assist in the development of solutions that fit the particular project and risk of concern.

**Table 6.15** *Example risk matrix – wind loading on buildings*

| Relevant climate change event | Confidence of prediction of change | Likelihood of climate change event | Construction process | Possible consequence of climate change | Significance of consequence on the construction activity | Overall risk of impact | Possible adaptation strategy | Residual risk following mitigation |
|---|---|---|---|---|---|---|---|---|
| Increase in number of depressions per season in winter (baseline of 5 per season) | Low | 63% | Design | Buildings two-storeys and under | Low | Low | Increase in average wind speed unlikely to affect structure | Low |
| | | | Design | Buildings 2–10 storeys | Low | Low | Increase in wind load should not require structural enhancement, but small increases in movements may occur | Low |
| | | | Design | Buildings 10–20 storeys | Moderate | Moderate | Increase in core walls to reduce deflection and resonance. Foundations may need commensurate stiffening | Low |
| | | | Design | Buildings above 20 storeys  Masts  Large bridges | High | High | Stabilise by increasing core walls; use of dynamic stabilisation; significant increase in foundations | Low |

## 6.8 POTENTIAL EFFECTS UPON ROOF DRAINAGE ARISING FROM CLIMATE CHANGE

### 6.8.1 Scope

**Identify the scope of the issue**

- is the key area of concern at the construction phase, during the design of the structure, or both?

Building roofs generally comprise hard surfaces of low permeability on which large quantities of water fall, requiring efficient drainage. Drainage is most often by passive gravity systems, wherein rainwater collects and is directed by drainage falls towards gulleys or other outlets. These gulleys and outlets are connected to downpipes that carry the water away to soakaways, main drainage systems or (rarely) direct discharge to watercourses. For larger areas, such as warehouse roofs, drainage is often by means of siphonic systems in which the drainage pipes are charged at pressure by outlets, thereby generating the necessary siphonic action.

## 6.8.2    Climate change context

**Identify the climate change issues**

- the climatic factors that may affect the construction process
- the climatic factors that may affect the design life of the building
- likelihood of a climate change event
- confidence in the predicted climate change scenarios

Predicted climate changes for the UK include the possibility of winter rainfall increasing both in volume and intensity. Current British standards are based on the climatic conditions that traditionally have been experienced. If average rainfall increases as predicted, many drains may become overloaded when subjected to increased volumes and intensity of rainfall.

Predicted climate change effects that have a bearing on roof drainage include the following:

- the annual volume of rainfall is not expected to change, but by the 2020s the UK can expect summer rainfall to decrease and winter rainfall to increase
- the average number of winter depressions is predicted to rise by 60 per cent from five to eight, increasing the potential for more driving rain (wind and rain) to be experienced.

Table 6.16 provides a summary of the extent of the predicted changes in climate conditions for the UK that are most relevant to roof drainage. For detailed climate change scenario information see Table 2.1.

**Table 6.16**    *Climate change predictions relevant to roof drainage*

| Factor | Parameter | 2020s | 2050s | 2080s |
|---|---|---|---|---|
| Precipitation | Average summer | 0–20% ↓ | 0–40% ↓ | 0–60% ↓ |
| | Average winter | 0–15% ↑ | 0–25% ↑ | 0–40% ↑ |
| | Extreme winter precipitation | | | 100% ↑* |
| Wind speed | Occurrence of lows (depressions below 1000 hPa) | | | ↑ in number of lows below 1000 hPa from 5 to 8 annually |

**Notes**

↓ = decrease;  ↑ = increase

\*    Medium-high emissions only

## 6.8.3    Current practice and guidance

**Identify appropriate guidance**

- British standards and codes of practice
- guidance to assist with the use of British standards
- general industry guidance, eg from BRE, CIRIA
- does the guidance take into account climate change?

All drainage that is currently installed, and which is in the process of being designed, is sized to carry the flows required by the current standards and codes of practice. Both the Building Regulations Part H and BS EN 12056-3:2000 refer designers to a table of intensities and durations. These have been established from statistical data available for current situations. Climate change is predicted to have the effect of increasing the intensities and possibly the durations. Current roof drainage systems have some capacity to retain short-term overloads, but have a finite maximum capacity. If the duration of heavy rainfall is extended by climate changes then leaks and overflows are possible, which could have deleterious effects on buildings. In the worst cases, lack of capacity could cause ponding on flat roofs, leading to leakage into the building.

In many countries where short-duration storms are normal and short-term intensities are very high, roof drainage is dealt with in a different way. Typically in tropical countries it is recognised that attempting to catch all the water and then carry it away through pipes may not be practical. The rainfall in these conditions is principally designed to spill off the roof through gargoyles or spouts, which are designed to throw the water away from the face of the buildings. The water is then collected at ground level and carried away. While, today, there are good reasons why such approaches would not be appropriate in the UK (splashing, ponding, lower evaporation rates etc), as the climate changes adaptation of overseas experience may become necessary.

**Existing roof drainage installations**

Very few commercial roofing systems have a guaranteed design life in excess of 20 years. This shorter design life has benefits in terms of the cost of addressing the roof drainage capacities in the existing commercial building stock. When the roof and/or its drainage system needs replacing the opportunity exists to carry out modification to reflect changing climatic requirements, for example by increasing the outlet capacities and resizing of downpipes.

It may also be possible to redesign and modify the roof layouts to increase the upstands around the roof and to incorporate overflow spouts. The former would add additional storage for the increased water volume until the existing down pipes could carry it away, while the latter would provide a safety valve to avoid overloading the roof.

For siphonic systems the situation would be more complicated, as a total redesign would need to be carried out to allow for increased capacity.

**New roof drainage installations**

For a new installation it is relatively inexpensive to increase the capacity of the drainage system through the use of additional outlets and larger-diameter downpipes. Consideration could also be given to the use of incomplete capture (as described above for tropical countries), but, as noted, there may be adverse public perception of this approach.

The specifications for external drainage at ground level have been increased after recent flooding events in the UK. Planning Policy Guidance Note 25 (DTLR, 2001), which relates exclusively to stormwater drainage, requires that new developments be designed for an increased storm flow, which is stated as being directly attributable to climate change. The guidance requires that flood risk should be evaluated for a 20 per cent increase in flows for the 1 per cent (100-year) and 0.5 per cent (200-year) storm probability. Currently, the building design codes and the Building Regulations for above ground drainage do not contain complementary advice, although it is suggested that, where available, local statistical rainfall data should be examined. The clear implication is that, for newly designed systems, an increase in capacity should be incorporated, arguably of the same order as that described in PPG25. This view might be tempered by the fact that below-ground drainage systems are generally more sensitive to intensity than to duration.

### 6.8.4 Risk matrix

Depending on the type of project, there are several options for reducing the risk of any climate-related issues associated with roof drainage. The following is not intended as a complete list of the options but rather provides examples that may help in the developing of solutions to fit particular projects and risks of concern. Table 6.16 identifies options for reducing the risk of rainfall-related events.

---

*Identify the climate change impacts on the project*

- what will be the consequence of the climate change impact?
- do I need to take account of climate change in the design of the structure?
- do I need to take account of climate change in the construction phase?
- are there any alternative solutions?

**Table 6.17**     *Risk matrix table – roof drainage*

| Relevant climate change event | Confidence of prediction of change | Likelihood of climate change event | Construction process | Possible consequence of climate change | Significance of consequence on the construction activity | Overall risk of impact | Possible adaptation strategy | Residual risk following mitigation |
|---|---|---|---|---|---|---|---|---|
| Wetter winters (66% wetter than average) | High | 7% | Design | Rainwater penetrating internal environment in domestic housing | Moderate | Low | Increase size of external drainage including gutters and downpipes | Low |
| | | | Design | Internal flooding of industrial premises from burst and overflowing siphonic pipes | High | Moderate | Increase size of downpipes and upstands to provide storage. Adaptation of existing systems may be difficult | Low |

# 7 Conclusions

This book has described the current state of predictions of climate changes and the potential consequences for selected building elements. The key messages are set out below.

It is realistic, based on current predictions, to assume that climate change will affect the construction industry. Climate predictions are made for three periods over the next century, known as the 2020s, 2050s and 2080s. It is especially for the later periods that some significant changes are predicted. As the design life of buildings and infrastructure can be in the order of 50–120 years it is clear that current designs will have to be able to withstand climate conditions within periods for which significant changes are predicted. The design and construction of buildings therefore needs to be adapted to cope with the predicted impacts of climate change.

Awareness of construction professionals of the possible impacts of climate change on the construction industry is low and should be increased. There is little evidence of UK construction companies implementing, or even considering, adaptation strategies or mitigation measures to reduce impacts from climate change. Possible reasons for this are given in this publication. Addressing these reasons should enable the industry to introduce appropriate adaptations into its decision-making processes.

Allowance is beginning to be made for climate change in the development of building standards and construction guidance. However, much of the current guidance lacks overt reference to climate change despite the likely effects falling within most project life-cycles. It is therefore difficult for a construction professional to know when climate change considerations should be applied to decision-making. Not only should current standards and guides specifically refer to the latest knowledge and understanding of climate change, but also publications should be progressively updated as knowledge increases.

The magnitude and timing of climate changes are uncertain. Their effects should therefore be addressed through risk-based processes. This publication has described a method for assessing the risk that should help designers and constructors make rational decisions about whether to incorporate climate change consequences in their projects. The method suggested follows the generic approach for risk decision-making and so can be adapted to fit with existing decision-making processes. It is, however, important to recognise that knowledge of climate change will expand as changes to our climate are seen and scientists' models are refined. The industry needs to be aware that the likelihood and magnitude of climate change events may change over the coming decades and that up-to-date figures should always be used in any risk assessment processes.

# APPENDICES

# A1     Questionnaire

The questionnaire below was distributed to professionals across several different sectors within the construction industry by URS, with the assistance of the steering group.

---

**YOUR DETAILS**

**QUESTION 1: Please enter your name, position and organisation details.** Although the results of the questionnaire **will remain anonymous**, internally it is important to know who has returned the questionnaire in order to categorise the risks.

Interviewer/contact . . . . . . . . . . . . . . . . . . . . . . . . . . . . . . . . . . . . . . . . . . . . . . . . . . . . . . . . . . . . . . . . . . . . . . . . . . . . . . . . . . . .

Respondent's name. . . . . . . . . . . . . . . . . . . . . . . . . . . . . . . . . . . . . . . . . . . . . . . . . . . . . . . . . . . . . . . . . . . . . . . . . . . . . . . . . . .

Position. . . . . . . . . . . . . . . . . . . . . . . . . . . . . . . . . . . . . . . . . . . . . . . . . . . . . . . . . . . . . . . . . . . . . . . . . . . . . . . . . . . . . . . . . . . . .

Years experience . . . . . . . . . . . . . . . . . . . . . . . . . . . . . . . . . . . . . . . . . . . . . . . . . . . . . . . . . . . . . . . . . . . . . . . . . . . . . . . . . . . . .

Organisation. . . . . . . . . . . . . . . . . . . . . . . . . . . . . . . . . . . . . . . . . . . . . . . . . . . . . . . . . . . . . . . . . . . . . . . . . . . . . . . . . . . . . . . . .

Date. . . . . . . . . . . . . . . . . . . . . . . . . . . . . . . . . . . . . . . . . . . . . . . . . . . . . . . . . . . . . . . . . . . . . . . . . . . . . . . . . . . . . . . . . . . . . . . .

---

**YOUR BUSINESS**

**QUESTION 2: Please tick <u>one</u> relevant box for each A, B and C.** This information will be used to categorise your business within the construction industry. If you wish to respond on behalf of more than one area of the construction industry (eg construction and use phase) please use a separate questionnaire.

---

**A. Please indicate from the list below which best reflects the business of your company:**

| | | | |
|---|---|---|---|
| ❑ | Insurer | ❑ | Supply chain |
| ❑ | Risk adviser | ❑ | Design |
| ❑ | Facilities management | ❑ | Housing |
| ❑ | Contract management | ❑ | Engineers |
| ❑ | Architect | ❑ | Other. . . . . . . . . . . . . . . . . . . . . . . . . . . . . . . . . |
| ❑ | Conveyance | | |

---

**B. Please indicate which construction sector is the main focus for your company:**

| | | | |
|---|---|---|---|
| ❑ | Non-domestic buildings (commercial/industrial) | ❑ | Transport infrastructure (roads, bridges etc) |
| ❑ | Domestic housing | ❑ | Other. . . . . . . . . . . . . . . . . . . . . . . . . . . . . . . . . |

---

**C. Which phase of construction is most applicable to your business?**

| | | | |
|---|---|---|---|
| ❑ | Design | ❑ | Use phase (facilities management) |
| ❑ | Construction | ❑ | Other. . . . . . . . . . . . . . . . . . . . . . . . . . . . . . . . . |

---

**QUESTION 3: Had you considered impacts of climate change on the construction industry prior to this questionnaire?**

| | | | |
|---|---|---|---|
| ❑ | Yes | ❑ | No |

---

## QUESTION 4: RISK IDENTIFICATION

**RISK**

**Please enter 8 climate change risks to your business.** The risks should be in order of **1 (greatest concern/problem)** to **8 (lesser concern)**. Please provide a precise description of the risk and think as broadly as possible. It may help to read the example and look at a selection of risks identified in Attachment A. **Please include and identify risks that have previously occurred, are a current problem as well as those that could also happen in the future** (keeping in mind the background information from Page 1). **Please note, flood risks are excluded from this project.**

**CONTROL**

Please include in this column the relevant management controls and counter measures that are **currently** in place in the organisation or **could be developed**. These could be organisational strategies, monitoring programmes, research & development, as well as financial measures, eg insurance.

| | RISK | TYPE OF BUILD<br>N = new<br>E = existing | CONTROL MEASURES |
|---|---|---|---|
| EG | *Subsidence from increased temperature causing shrinkage of clay soils leading to movement in foundations causing internal cracks and/or structural damage requiring subsequent repair.* | N | *For new domestic building stock, foundation depth has been/should be increased by x metres or x per cent* |
| 1 | | | |
| 2 | | | |
| 3 | | | |
| 4 | | | |
| 5 | | | |
| 6 | | | |
| 7 | | | |
| 8 | | | |

**PLEASE USE ADDITIONAL PAPER IF REQUIRED TO DESCRIBE THE RISKS AND ASSOCIATED CONTROL MEASURES**

## YOUR EXPERIENCES

We are interested in obtaining examples of **good practice case studies**.

**QUESTION 5:** Please identify any examples of good practice solutions that you have been involved with or know about through your general experiences. Please describe in relation to the following – **type and purpose of construction, design solution, cost savings and benefit of project.**

Eg Coastal erosion – A piled retaining wall is to be constructed along two sections of the Isle of Wight's Military Road to prevent it collapsing into the sea. A series of 22 m-long and 750 mm-diameter reinforced concrete piles will be capped by a tie beam and anchored back to a parallel row of piles further inland. Even when exposed, the piled wall will support the road allowing 4500 vehicles a day that use the road in tourist season to travel safely. Instruments will raise the alarm if there is any movement. The wall has a design life of 50 years.*

.................................................................................................................

.................................................................................................................

.................................................................................................................

.................................................................................................................

.................................................................................................................

.................................................................................................................

## WHAT TYPE OF GUIDANCE WOULD BE USEFUL?

**QUESTION 6:** Do you use external information in the form of publications to assist you with your business decisions? Please describe.

❑    Yes                              ❑    No

If yes, please describe below the type of publications you find most useful. It would help if you identified specific publications.

a.        Type – government publications, trade association guidance etc, and

b.        Format/structure – booklets, leaflets, CDs, seminars etc.

.................................................................................................................

.................................................................................................................

.................................................................................................................

.................................................................................................................

.................................................................................................................

## FURTHER COMMENTS

**QUESTION 7:** Do you have any other comments regarding this questionnaire or study?

.................................................................................................................

.................................................................................................................

.................................................................................................................

.................................................................................................................

.................................................................................................................

.................................................................................................................

ATTACHMENT A

For Question 4.

The risks identified in the list below are provided as examples of the potential risks to the construction industry from climate change. **The risks in the list below are only provided as examples and are <u>not</u> exhaustive.**

POTENTIAL RISKS OF CLIMATE CHANGE TO THE CONSTRUCTION INDUSTRY

- Increased weathering from driving rain, increased wind speed and increased temperature (external fabric)
- Dampness (internal mould)
- Construction down-time caused by health and safety concerns (increased wind speed, increased temperatures, severe rain storms etc)
- Increased amount of water (guttering and local infrastructure design to cope with increased volume of water)
- Subsidence through waterlogging or shrinkage in clay soils (movement in foundation causing structural damage)
- Durability and performance of materials including timber, concrete, masonry, stone, metals and plastic (chemical attack, frost damage, degradation, reinforcement corrosion, salt crystallisation, seasonal movements)
- Coastal erosion (damage/collapse of roads and buildings)
- Dislodged tiles from roofs from increased wind speeds (domestic)
- Structural failure due to wind speed/subsidence etc causing bridge, building or road collapse
- Concrete curing more difficult in hot/dry weather
- Greater UV damage to stored material
- Internal discomfort to building users, increasing demand for air conditioning
- Landslides causing damage to infrastructure (roads etc)
- More severe storms could increase structural damage to buildings and transport infrastructure
- Increased rain penetration affecting building façades and internal structures
- More intense rainfall may lead to drainage systems being unable to cope (roof, sewer, carriageway etc).
- Higher temperatures may cause ground contaminants to become more active and consequently attack foundations
- Development requiring intensive water usage will need to consider if water requirements can be met due to anticipated lower summer rainfall
- Decrease of winter energy consumption
- Milder winters and increased rainfall could lead to damp and mould problems

\* "Outlook changeable", *New civil engineer*, 30 January 2003.

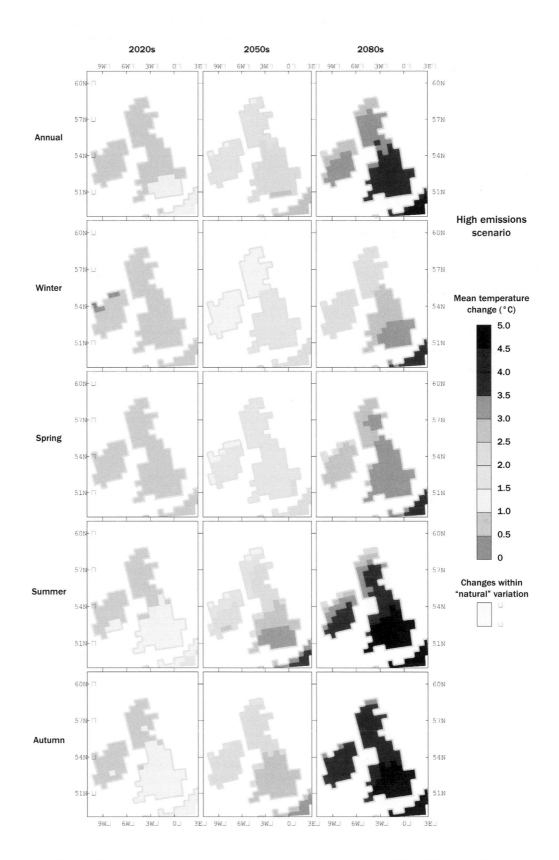

**Figure A2.1**     *Average seasonal and annual temperature for the 2020s, 2050s and 2080s for the high emissions scenario (Hulme et al, 2002)*

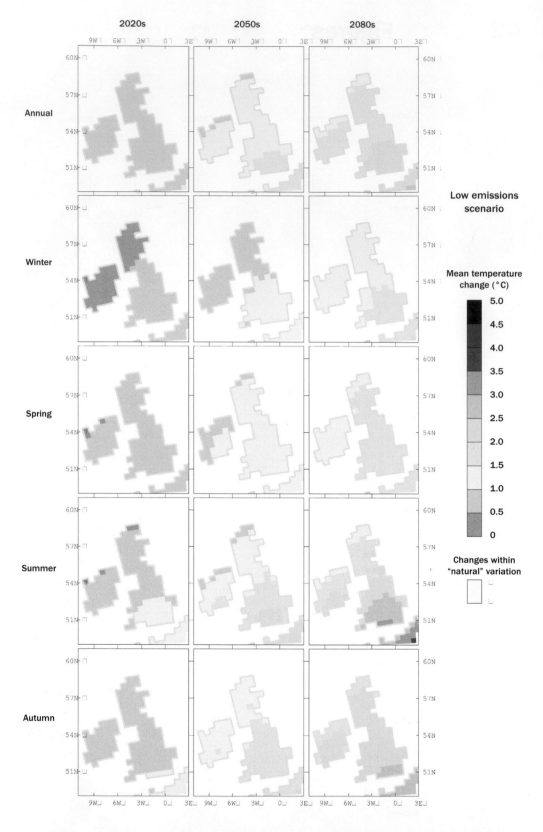

**Figure A2.2**    *Average seasonal and annual temperature for the 2020s, 2050s and 2080s for the low emissions scenario (Hulme et al, 2002)*

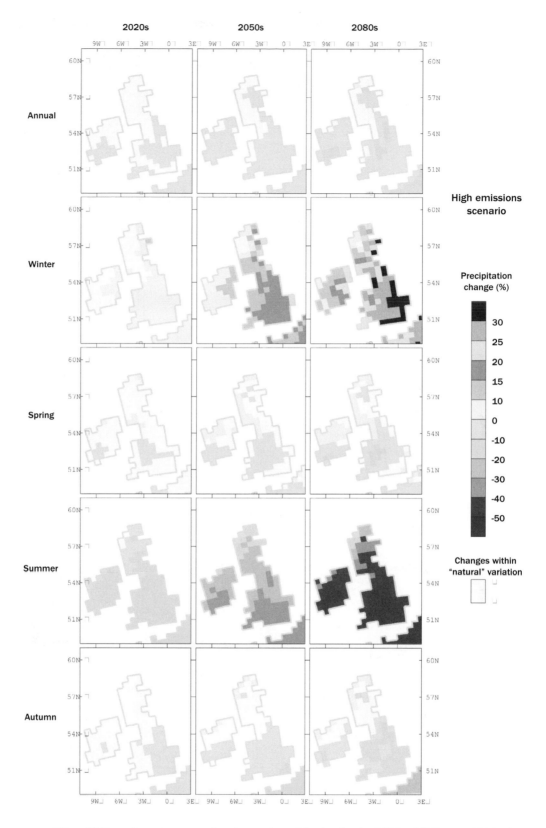

**Figure A2.3**    *Average seasonal and annual precipitation for the 2020s, 2050s and 2080s for the high emissions scenario (Hulme et al, 2002)*

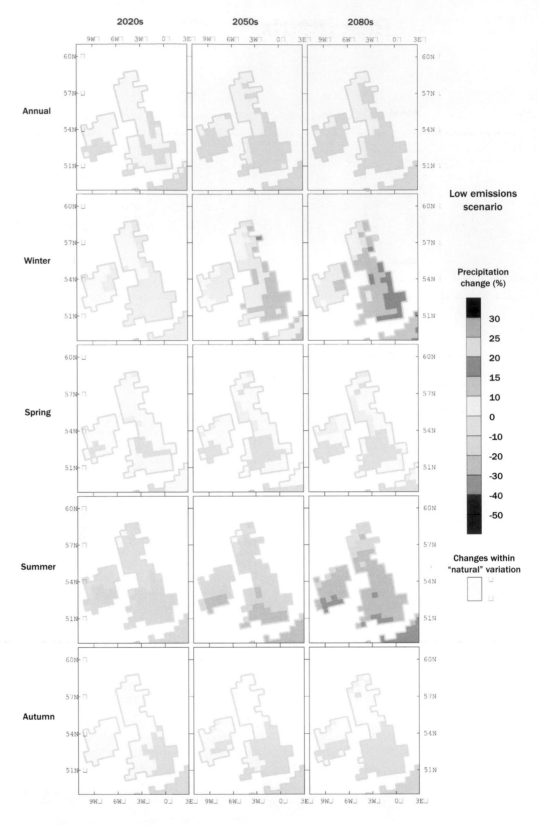

**Figure A2.4**    *Average seasonal and annual precipitation for the 2020s, 2050s and 2080s for the low emissions scenario (Hulme et al, 2002)*

# A3    Risk matrix

Table A3.1    *Example of risk matrix for the construction industry*

| Relevant climate change event | Confidence of prediction of change | Likelihood of climate change event | Construction process | Possible consequence of climate change | Significance of consequence on the construction activity | Overall risk of impact | Possible adaptation strategy | Residual risk following mitigation |
|---|---|---|---|---|---|---|---|---|
|  |  |  |  |  |  |  |  |  |
|  |  |  |  |  |  |  |  |  |
|  |  |  |  |  |  |  |  |  |
|  |  |  |  |  |  |  |  |  |
|  |  |  |  |  |  |  |  |  |
|  |  |  |  |  |  |  |  |  |
|  |  |  |  |  |  |  |  |  |

# A4 References

ABI (1999). *Subsidence: a global perspective*. General insurance research, Research Report no 1, Association of British Insurers, London

ABI (2003). *The vulnerability of UK property to windstorm damage*. Association of British Insurers, London

APMCA (1995). *Hot weather concreting*. Technical Bulletin 95/2, Australian Pre-Mixed Concrete Association, Sydney, Australia

Arup Research and Development (in press). *Climate change and the internal environment of buildings*. Technical memorandum, CIBSE, London

Beazant, G (2003). "Slippery slopes". *New Civil Engineer*, Jan

BRE (2002). *Assessing moisture in building materials. Part 1, Sources of moisture*. Good Repair Guide 33, Building Research Establishment, Garston

BRE (in press). *Mitigating the effects of climate change by roof design*. CIBSE, London

Building Regulations 2000. SI 2000/2531

Driscoll, R M C and Crilly, M S (2000). *Subsidence damage to domestic buildings: lessons learned and questions remaining*. Foundation for the Built Environment, London

DTLR (2001). *Development and flood risk*. Planning Policy Guidance Note 25 (PPG 25), HMSO, London

Ellis Jr, R D and Thomas, H R (2002). *The root causes of delays in highway construction*. London

English House Condition Survey (1991). <http://www.odpm.gov.uk/stellent/groups/odpm_housing/documents/page/odpm_house_603839.hcsp>. Accessed 16 Feb 2005

Eurocodes (2003). <http://www.eurocodes.co.uk>

Fawcett, W and Palmer, J (2004). *Good practice guidance for refurbishing occupied buildings*. C621, CIRIA, London

Graves, H M and Phillipson, M C (2000). *Potential implications of climate change in the built environment*. Foundation for the Built Environment Report 2, CRC, London

Greenwood, J R, Vickers, A W, Morgan, R P C, Coppin, N J and Norris, J E (2001). *Bioengineering – the Longham Wood cutting field trial*. Project Report 81, CIRIA, London

Hertin, J, Berkhout, F and Gann, D (2002). "Climate change and UK housebuilding sector: perceptions, impacts and adaptation". *Building Research and Information* no 30

Houghton, J T, Ding, Y, Griggs, D J, Noguer, M, van der Linden, P J, Dai, X, Maskell, K and Johnson, C A (eds) (2001). *Climate change 2001: the scientific basis. Contribution of Working Group I to the third assessment report of the Intergovernmental Panel on Climate Change*. Cambridge University Press, Cambridge

Hulme, M, Jenkins, G J, Lu, X, Turnpenny, J R, Mitchell, T D, Jones, R G, Lowe, J, Murphy, J M, Hassell, D, Boorman, P, McDonald, R and Hill, S (2002). *Climate change scenarios for the United Kingdom: the UKCIP02 scientific report*. Tyndall Centre for Climate Change Research, Norwich

Lowe, R (2001). *A review of recent and current initiatives on climate change and its impacts on the built environment: impact, effectiveness and recommendations*. Report no 10, Centre for Built Environment, Leeds

Marsh (2003). "Toxic moulds". Internal report, Marsh Ltd, London

McCarthy, J J, Canziani, O F, Leary, N A, Dokken, D J and White, K A (eds) (2001). *Climate change 2001: impacts, adaptation and vulnerability. Contribution of Working Group II to the third assessment report of the Intergovernmental Panel on Climate Change*. Cambridge University Press, Cambridge

Met Office (2003). <http://www.met-office.gov.uk/climate/uk/2003/>

NHBC (2003a). <http://www.nhbc.co.uk>

NHBC (2003b). "Building near trees". Chapter 4.2, *NHBC standards*, National House Building Council, London

Perry, J G (1994). *A guide to the management of building refurbishment*. Report 133, CIRIA, London

Perry, J and Brady, K (2003). *Infrastructure cuttings – condition appraisal and remedial treatment*. C591, CIRIA, London

Perry, J, Pedley, M and Reid, M (2003). *Infrastructure embankments – condition appraisal and remedial treatment*. C592, CIRIA, London

Price, D J and McInally, G (2001). *Climate change: review of levels of protection offered by flood prevention schemes*. Scottish Executive Central Research Unit, Edinburgh

Supreme Court of Queensland (1992). <www.courts.qld.gov.au>

UK Home Insurance (2003). <http://www.contents-insurance-uk.org.uk/flood-cover.htm>

Willows, R I and Connell, R K (eds) (2003). *Climate adaptation: risk uncertainty and decision-making. Technical Report*. UKCIP, Oxford

## British and European standards

BS 6399-2:1997 *Loading for buildings. Code of practice for wind loads*

BS 8104:1992 *Code of practice for assessing exposure of walls to wind driven rain*

BS EN 1990:2002 *Eurocode. Basis of structural design*

BS EN 12056-3:2000 *Gravity drainage systems inside buildings. Roof drainage, layout and calculation*

CP 3:Chapter V-2:1972 *Code of basic data for the design of buildings. Loading. Wind loads* (superseded)

# A5    Further reading

Arnell, N W (1999). "Climate change and water resources in Britain". *Climate change*, vol 39, no 1, pp 83–110

Arnell, N W (2004). "Climate-change impacts on river flows in Britain. The UKCIP02 scenarios". *The Journal*, vol 18, pp 112–117

Arnell, N W, and Reynard, N S (1996). "The effects of climate change due to global warming on river flows in Great Britain". *Journal of Hydrology*, vol 183, no 3–4, p 397

BRE (1996). *Low-rise buildings on shrinkable clay soils. Parts 1 and 2*. Digests 240 and 241, CRC, Garston

BRE (1999). *Wind loading on buildings. BS6399-2:1997 worked examples. Part 2: effective wind speeds for a site, and loads on a two storey house*. Digest 436, Building Research Establishment, Garston

BRE (1999). *Wind loading on buildings. BS6399-2:1997 worked examples. Part 3: loads on a portal frame building and on an office tower on a podium*. Digest 436, Building Research Establishment, Garston

BRE (1999). *Low-rise building foundations: the influence of trees in clay soils*. Digest 298, CRC, Garston

Building Regulations 2000 (SI 2000/2531). Approved documents:

C – *Site preparation and resistance to moisture* (2000)

F1 – *Means of ventilation* (2000)

F2 – *Condensation in roofs* (2000)

L1 – *Conservation of fuel and power in dwellings* (2002)

L2 – *Conservation of fuel and power in buildings other than dwellings* (2002)

CIRIA (1984). *The CIRIA guide to concrete construction in the Gulf region*. Special Publication 31, CIRIA, London

Cook, N and Narayanan, R S (1999). *Wind loading on buildings. Part 1: brief guidance for using BS6399-2:1997*. Digest 436, Building Research Establishment, Garston

Courtney, R, Chow, D and Levermore, G (2002). *Prioritising climate change issues with respect to other construction related agendas*. CRISP Commission 01/14 UMIST, Manchester

Crichton, D C (2001). *The implications of climate change for the insurance industry: an update and outlook to 2020*. Building Research Establishment, Garston

Defra (2003). *The impacts of climate change: implications for DEFRA*. Department for Environment, Food and Rural Affairs, London

Dill, M (2000). *A review of testing for moisture in building elements*. C538, CIRIA, London

Dlugolecki, A (2000). *Climate change and insurance*. HMSO, London

Driscoll, R (1995). *Assessment of damage in low-rise buildings*. Digest 251, CRC, Garston

DTLR (2002). *Preparing for floods: interim guidance for improving the flood resistance of domestic and small business properties*. Department of Transport, Local Government and the Regions, London

DTLR (2002). *Development on unstable land – Annex 1*. Planning Policy Guidance Note 14 (PPG14), The Stationery Office, London

East Midlands Sustainable Development Round Table (2000). *The potential impacts of climate change in the East Midlands. Summary report*. Environment Agency, Solihull

Environmental Resource Management (ERM) (2000). *Potential UK adaptation strategies for climate change. Technical report*. Department of Environment, Transport and Regions, London

Garrett, J and Nowak, F (1991). *Tackling condensation – a guide to causes of, and remedies for, surface condensation in traditional houses*. Building Research Establishment, Garston

Garvin, S L, Phillipson, M C, Sanders, C H, Hayles, C S and Dow, G T (1998). *Impact of climate change on building*. CRC, Garston

Hayne, M, Micheal-Leiba, M, Gordon, D, Lacey, R and Granger, K (2003). Chapter 7, "Landslide risks". In: K Granger and M Hayne (eds), *Natural hazards and the risks they pose to south-east Queensland*, Australian Geological Survey Organisation and Bureau of Meteorology, Canberra

Housing Forum (2002a). *Homing in on excellence: a commentary on the use of offsite fabrication methods for the UK house building industry*. The Housing Forum, London

Hulme, M, Turnpenny, J and Jenkins, G (2002). *Climate change scenarios for the United Kingdom: the UKCIP02 briefing report*. Tyndall Centre for Climate Change Research, Norwich

IStructE (2000). *Subsidence of low-rise buildings*. Institution of Structural Engineers, London

Kerr, A, Shackley, S, Milne, R and Allen, S (1999). *Climate change: Scottish implications scoping study*. Scottish Executive Central Research Unit, Edinburgh

Kersey, J, Wilby, R, Fleming, P and Shackley, S (2000). *The potential impacts of climate change in the East Midlands. Technical report*. Environment Agency, Solihull

London Climate Change Partnership (2002). *London's warming. The impacts of climate change on London. Summary report*. LCCP, London

London Climate Change Partnership (2002). *London's warming. The impacts of climate change on London. Technical report*. LCCP, London

McKenzie Hedger, M, Gawith, M, Brown, I, Connell, R and Downing, T (eds) (2000). *Climate change: identifying the responses. The first three years of the UK Climate Impacts Programme. Technical report*. UKCIP and DETR, Oxford

Metz, B, Davidson, O, Swart, R and Pan, J (eds) (2001). *Climate change 2001: mitigation. Contribution of Working Group III to the third assessment report of the Intergovernmental Panel on Climate Change*. Cambridge University Press, Cambridge

National Assembly for Wales (2000). *Wales: changing climate, challenging choices. The impacts of climate change in Wales from 2000 to 2080. Summary report*. National Assembly for Wales, Cardiff

National Assembly for Wales (2000). *Wales: changing climate, challenging choices. The impacts of climate change in Wales from 2000 to 2080. Technical report*. National Assembly for Wales, Cardiff

Nebojsa Nakicenovic and Rob Swart (eds) (2000). *International emissions scenarios. Special report of the Intergovernmental Panel on Climate Change.* Cambridge University Press, Cambridge

Perry, J, Pedley, M and Reid, M (2003). *Infrastructure embankments – condition appraisal and remedial treatment*, 2nd edn. C592, CIRIA, London

Shackley, S, Wood, R, Hornung, M, Hulme, M, Handley, J, Davies, E and Walsh, M (1998). *Changing by degrees – the impacts of climate change in the north west of England. Technical overview.* UKCIP, Oxford

Stirling, C (2002). *Assessing moisture in building materials. Part 2: Measuring moisture content. Part 3: interpreting moisture data.* Good Repair Guide 33, Building Research Establishment, Garston

Sustainability NorthWest (1998). *Everybody has an impact – climate change impacts in the north west of England. Summary report.* Sustainability NorthWest, Manchester

SWCCIP (2003). *Warming to the idea: meeting the challenge of climate change in the South West. Summary report.* South West Climate Change Impact Partnerships, <www.our-southwest.com/climate/scopingstudy.htm>

UKCIP (2002). *Warming up the region. Summary report. Yorkshire and Humber climate change impacts scoping study.* UKCIP, Oxford

UKCIP (nd) *Implications of climate change for Northern Ireland: informing strategy development.* UKCIP, Oxford

UKWIR (2003). *Effect of climate change on river flows and groundwater recharge. UKCIP02 scenarios.* 03/CL/04/2, United Kingdom Water Industry Research, London

Wade, S, Hossell, J, Hough, M, and Fenn, C (eds) (1999). *The impacts of climate change in the South East. Technical report.* WS Atkins, Epsom

Walker, M (ed) (2002). *Guide to the construction of reinforced concrete in the Arabian peninsula.* C577, CIRIA, London

Watson, R T *et al* (eds) (2001). *Climate change 2001: a synthesis report. A contribution of Working Groups I, II and III to the third assessment report of the Intergovernmental Panel on Climate Change.* Cambridge University Press, Cambridge

Wilson, M I and Burtwell, M H (2002). *Prioritising future construction research and adapting to climate change: infrastructure (transport and utilities).* CRISP, Commission 01/13, TRL Limited, Wokingham

WS Atkins, ADAS and the Meteorological Office (1999). *Rising to the challenge – impacts of climate change in the South East in the 21st century. Summary report.* WS Atkins, Epsom

Young, D K and Hooker, D P (1994). *A review of the geotechnical aspects for motorway widening works.* TRL Project Report 64, Highways Agency, London

### British and European standards

BS 8004:1986 *Code of practice for foundations*

BS 8500-1:2002 *Concrete. Complementary British Standard to BS EN 206-1 Part 1: Method of specifying and guidance for the specifier*

BS 8500-2:2002 *Concrete. Complementary British Standard to BS EN 206-1 Part 2: Specification for constituent materials and concrete*

CIRIA C638